FOOD AND BEVERAGE CONSUMPTION AND HEALTH

THE PINEAPPLE

PRODUCTION, UTILIZATION AND NUTRITIONAL PROPERTIES

FOOD AND BEVERAGE CONSUMPTION AND HEALTH

Additional books and e-books in this series can be found on Nova's website under the Series tab.

FOOD AND BEVERAGE CONSUMPTION AND HEALTH

THE PINEAPPLE

PRODUCTION, UTILIZATION AND NUTRITIONAL PROPERTIES

LYDIA HAMPTON
EDITOR

Library of Congress Cataloging-in-Publication Data

ISBN: 978-1-53614-594-6

Published by Nova Science Publishers, Inc. † New York

CONTENTS

PREFACE

The opening review included in *The Pineapple: Production, Utilization and Nutritional Properties* discusses the fundamental and applied aspects related to production of aroma compounds in pineapple fruit. Using different isolation techniques (vacuum distillation, simultaneous distillation-extraction, solvent extraction, dynamic headspace and headspace solid phase-microextraction), the volatile compounds are analyzed in conjunction, mainly with gas chromatography-mass spectrometry.

Internal browning or blackheart is an important physiological disorder of fresh pineapple fruit occurring during low temperature storage and a limiting factor during export. The authors discuss how phytosanitary treatments including heat, cold, or ionizing radiation can be used to control quarantine pests for market access.

Various classes of bioactive metabolites from pineapple reported in the scientific literature (and their nutritional properties) are discussed. The effects of these metabolites on human health are vital, and thus it is extremely important to understand their nutraceutical potential and future applications.

Lastly, the authors propose that the incorporation of pineapple peel flour into cooked meat products enhances moisture and texture, and improves probiotic survival during storage. Similarly, the antioxidant capacity of pineapple peel flour enhances lipid rancidity in cooked meat products.

Chapter 1 - This review discusses the fundamental and applied aspects related to production of aroma compounds in pineapple fruit. Using different isolation techniques (vacuum distillation, simultaneous distillation-extraction, solvent extraction, dynamic headspace and headspace solid phase-microextraction), the volatile compounds have been analyzed in conjunction, mainly, with gas chromatography-mass spectrometry. In pineapple fruit, over 400 volatile constituents have been reported, but only a few are considered important contributors to flavor. Aroma compounds in pineapple fruit have been reported to be influenced by various factors, including ripening season, maturity stage, varieties, and processing.

Chapter 2 - Internal browning or blackheart is an important physiological disorder of fresh pineapple fruit occurring during low temperature storage and a limiting factor during export. Phytosanitary treatments including heat, cold, or ionizing radiation can be used to control quarantine pests for market access. Irradiation is a viable alternative for insect disinfestation and can prolong the storage life of fresh fruits and vegetables. The effects of gamma irradiation on the quality of fresh produces may vary with species and cultivar, maturity stage, harvest season and radiation dose. In pineapple fruit, irradiation can increase the severity of internal browning symptoms beyond the effects of cold storage alone. Pre and postharvest treatments and conditions (maturity stage and harvesting season) are important factors that may be manipulated to reduce the severity of internal browning in irradiated cold-stored pineapples and postharvest losses. Certain fruit coatings also may help reduce internal browning symptoms and need further investigation.

Chapter 3 - Around the world, consumers are becoming much more aware of the potential health benefits of tropical fruits in human nutrition, as they are considered to be rich sources of bioactive metabolites along with their pleasant and strong aroma, appearance and taste. Pineapple is a native fruit of South America and highly appreciated for its exotic characteristics and presence of vital fruit nutrients. The nutritional and therapeutic values of this fruit are highly explored by biomedical researchers living in the South American region, which focus great interest in pineapple as a rich source of carotenoids, flavonoids, phenolic compounds and vitamin C with potential application on the reduced risk of several diseases, such as cancer, inflammation, cardiovascular, cataracts, diabetes, Alzheimer's disease, macular degeneration and neurodegenerative diseases. In addition, bromelain, a proteolytic enzyme which is present in pineapple possesses enormous pharmaceutical applications, including in anti-inflammatory, anti-diarrheal, anticancer, and inhibition of platelet aggregation. This chapter discusses various classes of bioactive metabolites from pineapple reported in the scientific literature and their nutritional properties. The effects of these metabolites on human health are vital, and hence it is extremely important to understand their nutraceutical potential and future applications. Thus, the chapter will also present the importance of pineapple fruit consumption in human diet due to the presence of important nutrients in this fruit.

Chapter 4 - Pineapple is one of the major tropical fruits produced in tropical and subtropical regions and consumed worldwide. As a consequence, during processing the peel represents a source of organic waste that can be employed as an important source of added value functional food ingredients. Pineapple peel flour contains a high amount of soluble and insoluble fiber, as well as polyphenols with antioxidant activity. The incorporation of pineapple peel flour into cooked meat products enhanced moisture and texture, and improved probiotic survival during storage. Similarly, the antioxidant capacity of pineapple peel flour enhanced lipid rancidity in cooked meat products. Pineapple co-products are low-cost, have low caloric content and can be employed as a fiber source in many food products.

In: The Pineapple
Editor: Lydia Hampton
ISBN: 978-1-53614-594-6

Chapter 1

PINEAPPLE FRUIT AROMA COMPOUNDS: STATE OF THE ART RESEARCH

Jorge A. Pino*
Food Industry Research Institute, La Habana, Cuba

ABSTRACT

This review discusses the fundamental and applied aspects related to production of aroma compounds in pineapple fruit. Using different isolation techniques (vacuum distillation, simultaneous distillation-extraction, solvent extraction, dynamic headspace and headspace solid phase-microextraction), the volatile compounds have been analyzed in conjunction, mainly, with gas chromatography-mass spectrometry. In pineapple fruit, over 400 volatile constituents have been reported, but only a few are considered important contributors to flavor. Aroma compounds in pineapple fruit have been reported to be influenced by various factors, including ripening season, maturity stage, varieties, and processing.

* Corresponding Author Email: jpino@iiia.edu.cu.

Keywords: pineapple, volatile compounds, analysis, isolation methods, aroma production

INTRODUCTION

Pineapple (*Ananas comosus* [L.] Merrill), named 'the king of fruits' because of its crown of leaves, is the most important representative of the Bromeliaceae family. It is believed that the species originated in southern Brazil and Paraguay, and was spread by the Amerindians to other parts of South and Central America [1].

Pineapple grows in several tropical and subtropical regions for local consumption and international export. Its economic importance is continuously increasing, along with efficient preservation and transportation [2]. According to FAO 2016 statistical data, Costa Rica, Brazil, Philippines, China, India, Thailand, Nigeria and Indonesia are the leading producing countries [3].

This is the third most important tropical fruit in the world, after banana and citrus [2], mainly because of its unique and impressive characteristics, as well as the refreshing sugar–acid balance.

Pineapple varieties are plentiful and approximately 100 varieties are now known worldwide, but only a few leading types are commercially available [2]. According to Morton [5], in international trade the many cultivars are grouped in four main classes: 'Smooth Cayenne' or 'Cayenne,' 'Red Spanish,' 'Queen,' and 'Abacaxi,' even if the types vary widely within each class. More recently, Po and Po [4] reported four groups too, but namely: 'Cayenne' (main cultivars Smooth Cayenne, Hilo, Kew, Champaka and Sarawak), 'Queen' (Moris, Mauritius, MacGregor, Ripley Queen and Alexandra), 'Spanish' (Singapore Spanish, Ruby, Red Spanish, Masmerah, Gandul, Hybrid 36, Selangor Green, Nangka and Betik), and 'Pernambuco and Mordilona' (Perolela). The 'Cayenne' group is the most important, with more than 70% of pineapple grown in the world for fresh fruit consumption and canning. Pineapple taxonomy was recently revised and simplified to two species (*Ananas comosus* (L.) Merrill and *Ananas macrodontes* Morren) and five botanical varieties [1].

Although pineapple is consumed fresh in producing countries, in the United States and Europe consumption of processed products as juices and canned fruit exceeds that of fresh pineapples, but demand for most canned products have been declining as consumers have shown increased preference for fresh and other processed-fruit products like juice and dried products.

Consumption of tropical fruits is mostly due to their unique and exotic flavor, while their nutritional value plays a minor role. Nevertheless, pineapple is an important source of potassium, magnesium, vitamins (C and A) and dietary fiber. Nutritionally, the fresh fruit is characterized by an energy content of 48.0-62.8 kcal, 12.6-14.8% total carbohydrates, 1.57-10.32% total sugars, 0.39-0.55% proteins, 0.11-0.17% total fat, 36.2-56.4 mg% vitamin C, 56-57 IU vitamin A, 52.5-115 mg% potassium, 13.0-14.4 mg% calcium, 12.0-20.0 mg% magnesium, 1.0-6.9 mg% sodium and 0.3-0.7 mg% iron [6].

Research work on pineapple aroma has been made for over 70 years in fresh fruit from different cultivars (not always specified) and processed foods. Until 2010, over 400 volatile constituents had been identified in this fruit and these results have been reviewed by numerous authors [7]; however, only some of them have been recognized as pineapple flavor contributors.

The aroma composition of pineapple fruit is summarized in this chapter, each section presenting an overview of important reports and a description of the characteristic features of the fruit's volatile compounds.

Analytical Methodologies Applied to Pineapple Aroma Research

Flavor is one of the key characteristics of food, one that highly impacts its consumption. It is very difficult to identify food flavor disregarding the aroma, this being the main reason why flavor research has largely meant studying the volatile compounds of a food. Sample preparation to ensure optimum performance is a crucial step in aroma analysis. The efficiency and selectivity of sample preparation techniques, as well as their applicability to relevant compounds and matrices, provide good results in several analytical techniques. Additionally, perfect sample preparation techniques are easy to use, low-cost, and compatible with many instrumental methods.

The analysis of aroma compounds has been the subject of important specialized treatises [8-15] and the present section will focus only on those methods applied in pineapple aroma research.

Several reviews on the chemical composition of pineapple aroma compounds were published earlier [16-20]. However, during the last decade, there has been substantial new data generated on aroma characterization of volatile compounds in pineapple by applying new extraction techniques.

The first step in aroma research is to select an adequate isolation technique preserving the sensory characteristics of the original product. Among the various instrumental techniques, gas chromatography (GC) is an effective and commonly used technique for aroma analysis. Aroma research had advanced enormously since the development of GC in the mid-1960s. Before the advent of GC, such research was particularly difficult. The introduction of mass spectrometry (MS) as a GC detector allows the combination of GC with mass spectrometry (GC-MS), instead of using a flame ionization detector (GC-FID) only. The GC-MS technique is effective and extremely valuable for obtaining structural information of aroma compounds by comparing fragments of target compounds with those of standard references from the database available in the GC-MS system. However, this technique is unable to determine the odor characteristics of the volatile compounds and their odor contribution to the sample. Therefore, it cannot identify the key aroma compounds in foods that affect its flavor, i.e., aroma-active or odor-active compounds, which are important for the acceptance of foods. Additionally, the concentration of some aroma compounds with high odor intensities is too low to be detected by GC-MS. Gas chromatography-olfactometry (GC-O), in combination with other analytical procedures has served as a very useful tool in flavor research for identifying and ranking the key odorants in various foods [21].

The techniques most commonly used for the isolation of volatile compounds from pineapple fruit and processed products included vacuum distillation [22-29], simultaneous distillation-extraction [31-33], solvent extraction [33-47], solvent assisted flavor evaporation [48], headspace analysis [31,48-52] and headspace solid-phase microextraction [7, 33, 53-69].

As no universal isolation method exists, it is essential to choose a method yielding an extract as representative as possible of the sensory properties of the fruit.

Distillation is a long-established isolation technique, and is widely used for separating components with different volatilities from non-volatile materials. Distillation using water vapor (steam distillation) or hydrodistillation (the material to be distilled comes in direct contact with boiling water) are used when the material to be distilled has a high boiling point and decomposition can occur if direct distillation is used. In both cases, the use of reduced pressure minimizes the possibility of decomposition during the isolation step. Nevertheless, hydrolysis of certain components, e.g., esters, and decomposition caused by high temperatures (greater than 60°C) are always present in this technique. Considering that water is the most abundant volatile component of the fruit, a second step based on solvent extraction is necessary to eliminate water from the distillate for subsequent analysis. Simultaneous distillation-extraction is included in this type of isolation technique. The acceptance of this technique comes from the fact that volatile compounds with medium to high boiling points are well recovered as a liquid isolate that it is sufficiently concentrated for GC analysis [70]. Solvents used for the extraction step in pineapple studies are ethyl ether and hexane.

Volatile compounds tend to be more soluble in organic solvents than in an aqueous solution and, therefore, solvent extraction may be used as a procedure to concentrate the aroma compounds. This isolation technique is inevitably time consuming and labor intensive, requiring the use of large amounts of an organic solvent which is seldom sufficiently pure to be directly used in aroma isolation. After extraction, the process to eliminate the solvent from the final extract is another difficulty. Solvent extraction is the second most used isolation technique in pineapple research. Solvents used for extraction in pineapple studies include ethyl ether, dichloromethane and pentane/dichloromethane (2:1 or 1:1).

Solvent assisted flavor evaporation (SAFE) was developed by Engel et al. [71]. It allows isolation of the volatile fraction from the matrix diluted in water or solvent using high vacuum distillation at low temperature (normally <40°C), thus avoiding potential flavor modification due to the formation of volatile compounds when heating ('artifacts'). This method has been successfully applied for aroma analysis [21]. Curiously, there is only one report with this technique applied to pineapple [48].

Headspace analysis may be accomplished by a static or dynamic technique [21]. In the first procedure, the equilibrium headspace above a food is analyzed. It is simple and easily automated, but sensitivity is generally very low. This is the reason for its infrequent use. In dynamic headspace, the sample is purged with some inert gas which strips volatiles from the sample. The volatiles in the purge gas must then be captured from the gas stream. The volatiles may be trapped via a cryogenic, polymer, charcoal, or other suitable trapping system. This technique favors the isolation of compounds with the highest vapor pressure and therefore, a distortion of the aroma profile occurs. Materials used for trapping volatiles in pineapple studies included Tenax GC, Tenax 60/80 and Tenax GR.

Headspace-solid phase microextraction (HS-SPME) combines extraction and preconcentration in one step, which is conducted using a modified syringe-like device that utilizes a polymeric extraction phase [21]. After sampling, the HS-SPME device can be coupled easily to a GC system for analysis, which has been successfully applied to the analysis of aroma compounds in fruits. The enormous acceptance of SPME is due to its undoubted merits: ease of operation, short extraction time, solvent-free nature, possibility of full automation, and easy coupling with GC, all of which decrease sample contamination and loss of analytes. The method was standardized by the Working Group on Methods of Analysis of the International Organization of the Flavor Industry (IOFI) in 2010 [72]. HS-SPME is the most used isolation technique in pineapple aroma research. Among the different types of fibers used in the analysis of pineapple volatiles, the most commonly used are 65 μm PDMS/DVB, 100 μm PDMS, 75 μm CAR/PDMS and 50/30 μm DVB/CAR/PDMS.

The polymeric extraction phase of the SPME fiber has a key role in determining aroma extraction behavior, which is selected based on the type of compounds in the sample in relation to the analytes of interest. Currently, diverse stationary phases with different polarities, providing a wide range of selectivity, are commercially available. In one of the studies, different types of fibers (100 μm PDMS, 65 μm PDMS/DVB and 50/30 μm DVB/CAR/PDMS) were compared using the same time and temperature conditions [33]. The results showed that, although PDMS (nonpolar coating) was more sensitive to esters than PDMS/DVB (polar coating) and PDMS/DVB was more sensitive to alcohols than PDMS, the best overall extraction efficiency was obtained when DVB/CAR/PDMS coating was used. However, considering that the aim of the study was to extract the odorant compounds, the global odor trapped in the fibers was also checked and the DVB/CAR/PDMS was chosen as the best fiber in this regard, generating the most representative odor. In another report [66], the performance of four fibers (100 μm PDMS, 75 μm CAR/PDMS, 85 μm CAR/PDMS and 65 μm PDMS/DVB) was investigated by comparing the total ion chromatogram area solely, and the PDMS/DVB fiber was slightly more capable of extracting the aroma compounds present in pineapple.

CONSTITUENTS INFLUENCING PINEAPPLE FLAVOR

Over the past 73 years, many techniques were applied to recognize the aroma-contributing constituents in pineapple. Several of the earlier studies on pineapple composition were conducted before the GC technique became available. As a result, it is known that pineapple flavor is a blend of many volatile constituents that are present in small amounts and in complex mixtures with the nonvolatile compounds. Several aroma compounds have been reported from fresh fruit and processed products, but comparison among these reports is problematic; since different varieties and products have been investigated. Additionally, different isolation and instrumental techniques have been used making comparison very difficult.

It was not until the first half of the 20th century that the first attempts were made to identify the main aroma compounds in pineapple. Haagen-Smith et al. [22, 23] were the first to investigate the volatiles of the cv. Smooth Cayenne grown in Hawaii. Large amounts of fresh fruit and juice were subjected to vacuum distillation at 20 mm Hg and the volatile material condensed in many traps kept at different temperatures. The small amount of oil thus obtained was typical of pineapple flavor. Identification was based on chemical or physical properties of the original compounds or their hydrolysis products, by the preparation of appropriate derivatives, i.e., dinitrobenzoates for the alcohols, *p*-phenylphenacyl esters for the acids, 2,4-dinitrophenylhydrazones for the carbonyl compounds and other specific chemicals for sulfur compounds. Many other reactions were used to identify sulfur compounds. The second study, on the analysis of canned Malayan juice, was attributed to Gawler in 1962 [24]. Using paper chromatography, he could separate and identify amino acids, organic acids, sugars and volatile carbonyl compounds such as 2,4-dinotro-phenylhydrazone derivatives.

The first study using GC in fruits from Hawaii was accredited to Mori in 1964, but these results were not published, although Connell [25] provided a summary of Mori's results later and contributed with research work on fresh Australian pineapples. Sixteen compounds, mainly esters and alcohols, were identified by comparison of GC retention times with reference compounds on three different GC stationary phases.

In 1965, two important studies on Smooth Cayenne fruit from Hawaii harvested in the winter season [26, 34], allowed identification of 2,5-dimethyl-4-hydroxy-3(2*H*)-furanone by several spectroscopic methods. This furanone has an odor described as 'burnt pineapple.' Also, *p*-allylphenol and γ-hexalactone were identified, and the presence of methyl

3-methylthiopropanoate and ethyl 3-methylthiopropanoate was confirmed. In those studies, preparative GC was used to isolate major compounds.

Howard and Hoffman in 1967 [49] developed a rapid means of comparing the aroma profile of different syrups from canned Malayan pineapples rings by headspace GC. The identification of 17 volatile constituents was based on retention times in three columns and by treating thc samplc with selective reagents.

One year later, Creveling et al. [36] analyzed an ether extract from freshly harvested Smooth Cayenne pineapples by analytical GC, as well as preparative GC for subsequent spectroscopic analyses (nuclear magnetic resonance and infrared spectroscopy). Twelve compounds were identified, including three 3-hydroxyesters, three lactones and two acetoxyesters. Among the lactones, γ- and δ-octalactone were characterized by a coconut-like aroma. The authors claimed that although the odors of the isolated 3-hydroxyesters were rather repulsive, their intensities suggest that these esters were important to pineapple aroma. 3-Hydroxyacid esters are present in several tropical fruits [73]. 3-Hydroxyacid esters are formed as intermediates during *de novo* synthesis and β-oxidation of fatty acids, but both pathways lead to opposite enantiomers. S-(+)-3-Hydroxyacyl-CoA-esters result from stereospecific hydration of Δ 2,3-*trans*-enoyl-CoA during β-oxidation; R-(-)-3-hydroxyacid derivatives are produced by reduction of 3-ketoacyl-S-ACP during fatty acid biosynthesis. Both pathways may be operative in the production of chiral 3-hydroxyacids and 3-hydroxyacid esters in tropical fruits [74, 75]. The enantiomeric composition of various lactones, hydroxy and acetoxy esters occurring in pineapples has been reported by Tressl et al. [73, 74].

The volatile compounds of the essence extracted from concentrated juice of Smooth Cayenne fruits from Hawaii were analyzed by Flath and Forrey [27] with the use of open tubular GC-MS. The introduction of this analytical procedure allowed separation and identification of 44 compounds. Four previously identified components: ethyl lactate, 2,5-dimethyl-4-hydroxy-3(2*H*)-furanone, 5-hydroxymethylfurfural, and *p*-allylphenol, did not survive injection into the stainless steel open-tubular GC columns, so would not have been detected, even if present in the extract. Compounds were identified by MS, and their identification was verified in two ways: first, by running a batch sample mass spectrum of pure standards on the same mass spectrometer and comparing spectra; and second, by co-injecting an authentic sample with the mixture of volatiles to compare its retention behavior with that of the tentatively-identified peak. Unfortunately, the significance to pineapple aroma of the identified compounds was not evaluated in that study.

Näf-Muller and Willhalm [37] added a considerable number of new constituents to the previously identified list. In total, they identified 59 aroma compounds in a concentrate prepared from fresh ripe fruits grown in Ivory Coast.

During the investigation to examine if allyl hexanoate was of natural occurrence in pineapple, Nitz and Drawert [39] analyzed ~20 kg of fresh fruits from Ivory Coast and ~6 kg from Kenia. After solvent extraction the extract was fractionated and enriched by preparative GC. GC-MS analysis of the corresponding fraction showed that allyl hexanoate was present.

The aroma constituents of these fruits from Ivory Coast have been isolated under enzyme inhibition, enriched by solvent extraction and fractionated on silica gel [40]. Analysis of the nonpolar fraction by capillary GC-FID and GC-MS showed the presence of at least 20 sesquiterpene hydrocarbons. Seven of them, α-copaene, β-ylangene, α-patchoulene, γ-gurjunene, germacrene D, α-muurolene, and δ-cadinene, were identified by comparison with data of authentic samples and published data. Before this study, only γ-eudesmol, a sesquiterpene alcohol with a selinane skeleton was reported [37]. One of the minor compounds, α-patchoulene, seems to contribute to the strong fruity-spicy odor of the fraction.

Nineteen other volatiles, occurring mainly in trace amounts in these fruits from Ivory Coast, including four nonterpenoid hydrocarbons and many carboxylic esters, were isolated under enzyme inhibition, enriched by continuous solvent extraction, fractionated on silica gel, and identified by capillary GC-FID and GC-MS [41]. Among them, 1-(*E,Z*)-3,5-undecatriene and 1-(*E,Z,Z*)-3,5,8-undecatetraene may contribute to the typical pineapple flavor. These conjugated alkenes combine a fragrant odor with extremely low odor detection thresholds. The corresponding *E,E* and *E,E,Z* isomers are much less odorous (by factors of 10^6 and 10^4, respectively). Disintegration of the fruit tissue without enzyme inhibition causes a rapid decrease of all undecaenes.

Ohta et al. [30] isolated the volatiles from canned Philippine pineapple juice by vacuum steam distillation and analyzed them by GC-FID and GC-MS. They reported methyl 4-acetoxyhexanoate and various carboxylic acids.

The aroma constituents of fresh Smooth Cayenne pineapple crown, pulp and intact fruit from Hawaii were analyzed by capillary GC and GC-MS [31]. The fruit was sampled using dynamic headspace sampling and vacuum steam distillation-extraction. A total of 183 and 160 compounds were identified by dynamic headspace and vacuum steam distillation-extraction, respectively. Analyses showed that the crown contains C_6 aldehydes and alcohols, while the pulp and intact fruit are characterized by an assortment of esters, hydrocarbons, alcohols and carbonyl compounds. Odor unit values, defined as the ratio of concentration of the compound to its odor threshold, indicated that the following compounds are the most important contributors to fresh pineapple aroma: 2,5-dimethyl-4-hydroxy-3(2*H*)-furanone, methyl 2-methylbutanoate, ethyl 2-methylbutanoate, ethyl acetate, ethyl hexanoate, ethyl butanoate, ethyl 2-methylpropanoate, methyl hexanoate and methyl butanoate. The sulfur-containing ester 3-methylthiopropyl acetate was reported for the first time in pineapple. This compound bears a relationship to the major esters, methyl and ethyl 3-methylthiopropanoate.

In another study, Takeoka et al. [28] analyzed the essence from juice concentration in Hawaii. Volatile compounds were isolated by solvent extraction, fractionated on neutral alumina, and analyzed by high-resolution GC-FID and GC-MS. Twenty-six constituents were identified for the first time in pineapple, including the following sulfur compounds: methyl (*E*)-3-methylthio-2-propenoate, ethyl (*Z*)-3-methylthio-2-propenoate, ethyl (*E*)-3-methylthio-2-propenoate, ethyl (*Z*)-3-methylthio-2-propenoate, methyl 4-methylthiobutanoate, ethyl 4-methylthiobutanoate (tentatively) and dimethyl trisulfide. The odor detection thresholds of these compounds were determined in water. Since the concentrations of the newly identified sulfur compounds were below their odor thresholds, they probably had little, if any, contribution to the overall aroma. Among the identified compounds, ethyl S-(+)-2-methylbutanoate was detected. This enantiomer is a potent odorant, with an odor detection threshold of 6 ng/kg, making it, according to the authors, the second largest odor contributor to pineapple aroma, after furaneol.

Volatile constituents of green and ripe pineapples were isolated and identified by GC-FID and GC-MS [29]. the numbers of 144 and 127 volatile constituents were found in green and ripe pineapples, respectively. Among a total of 157 constituents identified, 50 were identified for the first time in pineapple. Esters constituted over 80% of total volatiles from both green and ripe fruits. Diastereoisomers of butane-2,3-diol diacetate were satisfactorily separated in a GC chiral column.

Aroma compounds of fresh pineapple fruits from French Polynesia were isolated by continuous solvent extraction and analyzed by GC-FID and GC-MS. One hundred and eighteen compounds were identified according to their retention times on two capillary columns and their mass spectra. The composition found comprised 7 hydrocarbons (3.3%), 9 sulfur compounds (10.3%), 42 esters (44.9%), 10 lactones (11.5%), 11 carbonyl compounds (4.7%), 14 acids (7.3%), 11 alcohols and phenols (3.8%) and 14 miscellaneous compounds (14.3%). Four compounds were found at levels above 1 mg/kg: methyl octanoate (1.49 mg/kg), methyl 3-methylthiopropanoate (1.14 mg/kg), methyl hexanoate (1.1 mg/kg) and 3-methyl-2,5-furanedione (1.07 mg/kg), while three were detected between 0.5 mg/kg and 1 mg/kg and 25 between 0.1 mg/kg and 0.5 mg/kg. Of these compounds, 47 were newly detected aroma compounds [43].

Preston et al. [44], from extracts prepared by solvent extraction of volatile compounds from self-prepared pineapple juices and commercial pineapple recovery aroma phases, used on-line capillary GC-isotope ratio MS in the combustion (C) and pyrolysis (P) modes (HRGC-C/P-IRMS) to determine the isotope ratio, expressed as per mil for $^{13}C/^{12}C$ ($\delta^{13}C_{VPDB}$) and $^{2}H/^{1}H$ ($\delta^{2}H_{VSMOW}$) values of selected pineapple aroma compounds. Fresh pineapple fruits from seven regions (Costa Rica, Ghana, Honduras, Ivory

Coast, Réunion, South Africa and Thailand) were purchased from fruit distributors or the industry. Authors concluded that the δ^2H analysis via on-line HRGC-P-IRMS analysis is a helpful tool in the authentication of food flavors.

As a wide hybridization program initiated by CIRAD-FLHOR in 1988 to improve overall pineapple quality, a selection program resulted in the isolation of a new cultivar (FLHORAN41). The female parent (i.e., the seed parent) was the cv. Smooth Cayenne. (clone 'HA 10'), a typical Hawaiian pineapple, and the male parent (i.e., the pollen parent) was the cv. Manzana (clone 'CO 24'), a variety grown for local consumption at high altitude, primarily in Colombia. Depending on storage conditions, the FLHORAN41 cultivar develops a red-orange to scarlet shell color at maturity, making this cultivar potentially quite attractive to the consumer. Furthermore, its flesh at maturity is golden yellow while the cv. Smooth Cayenne has a pale, yellowish flesh. The physicochemical characteristics, sugars, organic acids, carotenoids, anthocyanins, volatile compounds and cell wall polysaccharides of a new pineapple hybrid were determined throughout maturation and compared with the cv. Smooth Cayenne [45]. At full maturity, the cv. FLHORAN41 has a higher titratable acidity and soluble solids content than the cv. Smooth Cayenne. The golden yellow flesh and red-orange to scarlet shell of ripe hybrid fruits are due to carotenoid and anthocyanin levels that are, respectively, 2.5 and 1.5 times higher than those of the flesh and shell of ripe cv. Smooth Cayenne, respectively. Aroma components were isolated by means of solvent extraction and analyzed by GC-MS. Of the 65 peaks detected, 49 were identified, and these were mainly aliphatic, hydroxy, and acetoxy esters (~40-50% of the total concentration in ripe fruits) and terpenes. During maturation of the cv. FLHORAN41, there was an increase in all classes of volatiles (mainly terpene hydrocarbons and esters), although their relative proportions were similar in both cultivars at full maturity.

Free aroma constituents of the cv. Perolera grown in Colombia were obtained by continuous solvent extraction and analyzed by GC-FID, GC-MS and GC-O [46]. Sixty-seven constituents were identified, with methyl 2-methyl butanoate, methyl hexanoate, methyl 4-acetoxy-hexanoate, methyl 5-acetoxy-hexanoate and methyl 3-methylthiopropanoate as major components. Glycosidically-bound aroma compounds were also identified.

The flavor profile of juices made from fresh-cut pineapple fruits from eight regions was studied in comparison to that of pineapple recovery aroma phases, juice concentrates, as well as commercially available juices and jams were investigated [47]. Isolation of volatile compounds was made by continuous solvent extraction. GC-MS analysis of juices made from fresh-cut Super Sweet fruit revealed the presence of esters, with methyl 2-methylbutanoate, methyl 3-methylthiopropanoate, methyl butanoate, methyl hexanoate, ethyl hexanoate and ethyl 3-methylthiopropanoate, as well as 2,5-dimethyl-4-methoxy-3(2*H*)-furanone and 2,5-dimethyl-4-hydroxy-3(2*H*)-furanone as major volatile constituents. A corresponding flavor profile was rarely found in pineapple recovery aroma phases under study. In most cases, the characteristic methyl esters and hydroxy or acetoxy esters were lacking completely or appeared only in minor amounts in these products. Only a few of the commercial single strength juices revealed fruit-related flavor profiles, juices produced from concentrates mostly exhibited an aroma composition close to that of concentrates, i.e., they were predominantly determined by their contents of 2,5-dimethyl-4-methoxy-3(2*H*)-furanone and did not show a fruit-related ester distribution. Similarly, the jams studied were poor in typical pineapple volatiles.

By application of aroma extract dilution analysis (AEDA) to an aroma distillate prepared from fresh cv. Super Sweet (F-2000) using SAFE, 29 odor-active compounds were detected in a flavor dilution (FD) factor range of 2 to 4096 [48]. In addition, calculation of odor activity values (OAVs), defined as the ratio of concentration of the compound to its odor threshold, twelve selected odorants revealed five key odorants in fresh pineapple: 2,5-dimethyl-4-hydroxy-3(2*H*)-furanone (sweet, pineapple-like, caramel-like), ethyl 2-methylpropanoate (fruity), ethyl 2-methylbutanoate (fruity), methyl 2-methylbutanoate (fruity, apple-like) and 1-(*E,Z*)-3,5-undecatriene (fresh, pineapple-like). A mixture of these twelve odorants in amounts equal to those in the fresh fruit resulted in an odor like that of the fresh fruit. The results of omission tests using the model mixture showed that furanone and ethyl 2-methylbutanoate are odorants having character impact in pineapple.

Aroma components and characteristic odorants in headspace aroma of Philippine pineapple were studied by use of vacuum extraction with a passivated stainless-steel canister [51]. Among 56 compounds identified, 26 were aliphatic methyl esters, accounting for approximately 93% of the total volatiles, including methyl 2-methylbutanoate and methyl hexanoate as the major components. GC-O led to the detection of methyl 2-methylbutanoate, ethyl 2-methylbutanoate, acetaldehyde, 1-(*E,Z*)-3,5-undecatriene,methyl butanoate, and methyl (*E*)-3-hexenoate as potent odorants. Methyl 3-methylthiopropanoate, which is known to have a characteristic pineapple flavor, was not detected as a potent odorant due to its trace quantity.

Braga et al. [54] investigated the volatiles composition of fresh pineapples from Brazil and fruits dried under normal and modified atmosphere conditions by adding 0.5% v/v ethanol to the drying air in a lab-scale tunnel dryer. Volatile compounds were isolated by HS-SPME and analyzed by GC-MS. Important aroma compounds of pineapple were detected in fresh as well as dried samples. Most of the volatiles identified were esters, alcohols, aldehydes, hydrocarbons (including monoterpenes), and sulfur esters.

A comprehensive analysis of volatile constituents in Costa Rican Gold cultivar fruits by means of HS-SPME and GC-MS techniques was reported by Montero-Calderón et al. [56]. Physicochemical attributes, aroma profile, and odor contribution of pineapple flesh were studied for the top, middle, and bottom cross-sections cut along the central axis of the fruit. Relationships between volatile and nonvolatile compounds were also studied. Twenty volatiles were identified and quantified. Among them, esters were the major components, accounting for 90% of total extracted aroma. Methyl butanoate, methyl 2-methylbutanoate, and methyl hexanoate were the three most abundant constituents (74% of total volatiles). Most odor-active contributors were methyl 2-methylbutanoate, ethyl 2-methylbutanoate and 2,5-dimethyl 4-methoxy-3(2*H*)-furanone. Aroma profile did not vary along the fruit, but volatile compounds content significantly varied along the fruit, from 7.56 to 10.91 mg/kg, in the top and bottom cross-sections, respectively. In addition, the concentration of most odor-active compounds increased from the top to the bottom 3rd of the fruit, simultaneously with soluble solids content (SS) and titratable acidity (TA), differences attributed to fruitlets different degrees of ripening. Large changes in the SS/TA ratio and volatiles content throughout the fruit found in this study are likely to cause important differences among individual fresh-cut pineapple rings, compromising consumer perception and acceptance of the product. Such findings highlighted the need to include volatiles content and SS/TA ratio and their variability along the fruit as selection criteria for pineapples to be processed, as well as for quality assessment of the fresh-cut fruit.

Characteristic aroma compounds from different parts of the cv. Smooth Cayenne fruits grown in Yunnan province (China), were analyzed by HS-SPME and GC-MS [61]. The major volatiles were esters, terpenes, ketones and aldehydes. The number and content of compounds detected in pulp were higher than those found in the core. In pulp, the characteristic aroma compounds were ethyl 2-methylbutanoate, ethyl hexanoate, 2,5-dimethyl-4-hydroxy-3(2*H*)-furanone (DMHF), decanal, ethyl 3-methylthiopropionate, ethyl butanoate, and ethyl (*E*)-3-hexenoate; while in the core the main compounds were ethyl 2-methylbutanoate, ethyl hexanoate and DMHF. The highest OAVs were found to correspond to ethyl 2-methylbutanoate, followed by ethyl hexanoate and DMHF. The OAVs found in pulp were higher than those in the core.

A comprehensive analysis of aroma compounds in cv. Hawaii pineapples grown in Brazil was reported by Pedroso et al. [62]. Combining qualitative data from the chromatographic structure of two-dimensional GC-FID (GC x GC-FID) and that from GC-MS should result in a more accurate assignment of peak identities than simple analysis by GC-MS, where coelution of analytes is unavoidable in highly complex samples (rendering spectra unsuitable for qualitative purposes) or for compounds in very low concentrations. The use of data from GC x GC-FID combined with GC-MS can reveal coelutions that were not detected by mass spectra deconvolution software. In addition, some compounds can be identified according to the structure of the GC x GC-FID chromatogram. In this study, the volatile fractions of fresh and dehydrated pineapple pulp were evaluated. The extraction of volatiles was performed by HS-SPME. The results using both techniques were combined to improve compound identifications.

Pino [33] combined different isolation approaches, including HS-SPME, simultaneous distillation-extraction and solvent extraction in conjunction with GC-FID, GC-MS, AEDA and OAVs to analyze the volatiles from the Red Spanish cultivar, the most important in Cuba, and to estimate the most odor-active compounds. In total, 94 compounds were identified, 72 of them positively identified against pure standards. Twenty constituents were considered as odor-active compounds, of which ethyl 2-methylbutanoate, 2,5-dimethyl-4-hydroxy-3(2*H*)-furanone, 1-(*E,Z,Z*)-3,5,8-undecatetraene,ethyl 3-methylthiopropanoate, 1-(*E,Z*)-3,5-undecatriene, ethyl hexanoate and methyl hexanoate were the more relevant odor contributors with regard to the characteristic pineapple aroma, while the others are responsible for fruity and sweet odor notes.

Wei et al. [66] developed a HS-SPME method for pineapple aroma compounds. Different sample preparation procedures involving SPME fibers, including the addition of sodium chloride, extraction time, and temperature, were evaluated to optimize the method. For the optimized method, 6.5 g of ground pineapple flesh was placed in a 20-mL headspace vial with 1 g of sodium chloride; a 65 μm PDMS/DVB fiber was used for extraction at 50°C for 40 min with continuous stirring. The method was validated by determining its repeatability and recovery. Results demonstrated that this procedure is suitable as a simple, rapid, and solvent-free extraction technique for the analysis of volatiles. Using this method, 15 compounds were identified, and the amount was 1019.78 μg/kg in the cv. Shenwan, including seven esters, two aldehydes, two lactones, as well as, one terpene, ketone, alcohol and hydrocarbon. A tentative study to estimate the contribution of the identified compounds to the aroma of this cultivar, based on their OAVs, indicated four compounds that were characteristic aroma compounds, including methyl hexanoate, δ-octalactone, decanal and geranyl acetone.

Despite detailed information of the qualitative and quantitative composition of pineapple aroma compounds, solid information about the significance of individual constituents to the aroma and flavor has long been lacking. Therefore, recent research work has focused on the determination of aroma-active compounds [31, 33, 48]. Undoubtedly, a useful tool to identify the most odor-active compounds among the volatiles is the combination of AEDA [76] with calculation of OAVs, as well as sensory studies on model mixtures containing the identified aroma-active compounds in the same concentrations as determined in fresh pineapple.

At present, there are no doubts that 2,5-dimethyl-4-hydroxy-3(2*H*)-furanone, also named 'furaneol,' a relatively hydrophilic and not very stable substance, is an essential pineapple flavor compound. This furanone has also been identified in many fruits, including strawberries and mangoes [38]. The odor and taste threshold values in water are 0.1-0.2 mg/kg and 0.03 mg/kg, respectively [77]. At high concentrations its odor is strongly caramel-like, while concentrations near its odor threshold give a pleasant fruit odor, reminiscent of strawberry and pineapple [78].

The contribution to pineapple flavor of methyl and ethyl esters of 3-methylthiopropanoic acid is not clear. Preliminary taste tests were conducted with both thioesters added to stripped juice [35]. At low levels, they were reportedly acceptable, although no pineapple character was found, while at higher contents an 'overripe' taste was found. On the other hand, both thioesters were found to be odor-active compounds using OAV [31] and the combination of OAV and AEDA procedures [33]. Ethyl 3-methylthiopropanoate was characterized as 'pineapple-like' by GC-sniffing technique [33, 41]. In addition, they are included as pineapple-like aroma compounds at low concentrations [79].

Two minor nonterpenoid hydrocarbons,1-(*E,Z*)-3,5-undecatriene and 1-(*E,Z,Z*)-3,5,8-undecatetraene, are recognized as important contributors to fresh-cut pineapple aroma due to their low odor detection threshold [41].

Lactones are considered important for pineapple aroma, in particular γ-octalactone, δ-octalactone and γ-nonalactone, which are responsible for the coconut note in the aroma of fully-ripe pineapple.

Among the identified volatile compounds, aliphatic esters have also been considered important for pineapple aroma. Aliphatic esters often have fruity notes [79]. Berger [41] found that esters such as 2-methylbutanoates and hexanoates contribute fruity notes to fresh pineapple.

The studies reported by Tokitomo et al. [48] and Pino [33] corroborated those results.

GLYCOSIDICALLY-BOUND AROMA COMPOUNDS

The presence of glycosidically bound volatile compounds in fruits has been well established. These compounds can be released from the carbohydrate moiety by enzymatic or chemical cleavage during maturation, industrial pretreatment or processing, and can be considered aroma precursors [80].

Wu et al. [42] were the initiators of the analysis of free and glycosidically bound volatiles from pineapple juice by separation on an Amberlite XAD-2 column. Volatile compounds from bound fractions were released by almond β-glucosidase hydrolysis. By use of γ-valerolactone as internal standard, volatile components of free (4963 μg/kg) and bound (1246 μg/kg) fractions were determined by GC-FID and GC-MS. Glycosidically bound 2,5-dimethyl-4-hydroxy-3(2*H*)-furanone, phenols, lactones, alcohols, acids, and aldehydes were observed in pineapple for the first time. Phosphatase was also used to hydrolyze the bound fraction, but no volatile compounds were released. Glycosidically bound 2,5-dimethyl-4-hydroxy-3(2*H*)-furanone was further confirmed by HPLC analysis.

Glycosidically bound aroma compounds of the cv. Perolera grown in Colombia were identified by GC and GC-MS after isolation of the glycosidic fraction obtained by Amberlite XAD-2 adsorption and methanol elution followed by hydrolysis with a commercial enzyme [46]. In total, 17 bound volatile compounds (aglycones) were identified in an amount of 12.1 mg/kg pulp. Identified aglycones were mainly hydroxylated compounds and carboxylic acids, from which coniferyl alcohol, hexadecanoic acid, 2,5-dimethyl-4-hydroxy-3(2*H*)-furanone and 4-vinylguaiacol were the major ones.

CHANGES IN PINEAPPLE FLAVOR PROFILE

The aroma of pineapple fruit has been reported to be influenced by various factors, including ripening season, fruit maturity stage, varieties and processing conditions.

Influence of Ripening Season

The first attempt to evaluate the aroma compounds in pineapple fruits ripened in different seasons was made by Haagen-Smith et al. [22], pioneers in research concerning pineapple flavor. They investigated the volatile compounds of the cv. Smooth Cayenne grown in Hawaii in two different ripening seasons. Differences were found between summer and winter fruits in extraction yield and composition. Major constituents of summer fruits were ethanol and ethyl acetate, with smaller quantities of acetaldehyde, ethyl acrylate, ethyl 3-methylbutanoate, ethyl hexanoate, methyl and ethyl esters of C_5 unsaturated acid, methyl 3-methylthiopropanoate and acetic acid, while for winter fruits they found ethyl acetate as the major component, followed by acetaldehyde, methyl 3-methylbutanoate, methyl pentanoate, methyl 4-methylpentanoate, methyl octanoate and methyl 3-methylthiopropanoate.

Liu et al. [59] compared the aroma compounds, analyzed by HS-SPME and GC-MS, among cv. Shenwan fruits ripened in each season. Results showed the diversity in volatile compound composition among them, with a big range of aroma components, especially esters and heterocyclic compounds, the highest levels of characteristic aromas corresponding to summer fruits. This is the reason why summer fruits excel in fragrance and, consequently, consumers' preference. Ten components were detected in spring fruits, including four compounds found only in this season. Major constituents were methyl hexanoate, 1,3,7-octatriene-3,7-dimethyl and methyl octanoate, with relative contents of 38.94, 26.32 and 10.34%, respectively. Nineteen compounds were detected in summer fruits, with 10 unique components. The predominant ones were methyl

2-methylbutanoate, methyl hexanoate, and 2-hydroxy-N-(2-morpholinoethyl)-4-phenylbutanamide, with relative contents of 24.95, 24.69 and 9.60%. Eleven compounds were detected in autumn fruits, including three unique components, the main ones being methyl hexanoate, methyl 2-methylbutanoate and methyl butanoate, with relative contents of 57.67, 18.52 and 8.76%, respectively. Twelve components were detected in winter fruits, including four unique compounds. The predominant ones were methyl hexanoate, methyl octanoate and *cis*-β-ocimene with relative contents of 63.08, 15.82 and 7.78%, respectively. The relative contents of methyl 3-methylthiopropanoate in the fruits from each season were 0.89, 4.09, 0.45 and 0.99%, respectively. Compounds were only tentatively identified by comparing their mass spectra with those in a mass spectra library. Thus, some identifications are doubtful, e.g., 2-hydroxy-N-(2-morpholinoethyl)-4-phenylbutanamide, which is not a volatile compound.

Influence of Maturity Stage

Umano et al. [29] examined the volatile constituents of green and ripe pineapples (unknown variety) from Philippines. The numbers of volatiles found were 144 and 127 in green and ripe pineapples, respectively. Major constituents in green fruits were ethyl acetate and ethyl 3-methylthiopropanoate, while those in ripe fruits were ethyl acetate and butane-2,3-diol diacetate.

Steingass et al. [68] examined the quantitative chirospecific analysis of δ-lactones (octalactone and decalactone) and γ-lactones (hexalactone, octalactone and decalactone) using HS-SPME and enantioselective GC-selected ion monitoring MS (HS-SPME-GC-SIM-MS). Based on a stable isotope dilution assay, lactone profiles of three post-harvest stages of on-plant ripened air-freighted pineapples were compared to green-ripe sea-freighted fruits, thus covering the entire life cycles from the first day of fruits' availability in the European market (air-freight: 2 days after harvest (dah); sea-freight: 14 dah until the end of their commercial shelf-life, which was generally reached two weeks after further post-harvest storage.

Concentration of lactones and their enantiomeric distribution varied between sea- and air-freighted fruits, thus allowing the authentication of the pineapples according to their post-harvest procedures and fruit logistics. Fresh fruits harvested at full maturity were characterized by γ-hexalactone of high enantiomeric purity remaining stable during the whole post-harvest period. In contrast, the enantiomeric purity of γ-hexalactone significantly decreased during post-harvest storage of sea-freighted pineapples. The biogenetical background and the potential of chirospecific analysis of lactones for authentication and quality evaluation of fresh pineapple fruits were discussed. The authors claimed that enantioselective analysis of lactones may be successfully applied for the authentication of air-freighted pineapples traded as a premium niche product in the higher priced segment, thus allowing the substantiation of consumers' deception and fraud by misleading advertisement of the fruits.

Qualitative ripening-dependent changes of the cv. Extra Sweet aroma compounds were analyzed by HS-SPME and comprehensive two-dimensional GC-quadrupole MS (HS-SPME-GC × GC-qMS) [7, 69]. Early green-ripe stage, post-harvest ripened, and green-ripe fruits at the end of their commercial shelf-life were compared to air-freighted pineapples harvested at full maturity. More than 290 volatiles were identified, most of them comprised esters (methyl and ethyl esters of saturated and unsaturated fatty acids), terpenes, alcohols, aldehydes, ketones, free fatty acids, and many γ- and δ-lactones. The structured separation space obtained by GC × GC allowed revealing various homologous series of compound classes as well as clustering of sesquiterpenes. Post-harvest ripening increased the diversity of volatile profiles compared to both early green-ripe maturity stages and on-plant ripened fruits. Profile patterns presented in the contour plots were evaluated applying image-processing techniques and subsequent multivariate statistical data analysis. Statistical methods comprised unsupervised hierarchical cluster analysis (HCA) and principal component analysis (PCA) to classify the samples. Supervised partial least squares discriminant analysis (PLS-DA) and partial least squares (PLS) regression were applied to discriminate different ripening stages and describe the development of volatile compounds during postharvest storage, respectively. The workflow permitted a rapid distinction between premature green-ripe pineapples and postharvest-ripened sea-freighted fruits. Volatile profiles of fully ripe air-freighted fruits were like those of green-ripe fruits postharvest-ripened for six days after simulated sea-freight export, after PCA with only two principal components. However, PCA considering also the third principal component allowed differentiation between air-freighted fruits and the four successive postharvest maturity stages of sea-freighted pineapples.

Influence of Varieties

Aroma compounds from two pineapple cultivars (Tainong No. 4 and No. 6) were isolated by HS-SPME, identified and quantified by GC-MS [63]. In Tainong No. 4 and No. 6 fruits, a total of 11 and 28 volatile compounds were quantified, with total concentrations of 1080.44 μg/kg and 380.66 μg/kg in Tainong No. 4 and No. 6 cultivars, respectively. The OAVs of volatile compounds were also calculated. According to the OAVs, four of them were defined as the characteristic aroma compounds for the cv. Tainong No. 4, including furaneol, methyl 3-methylthiopropanoate, ethyl 3-methylthiopropanoate and δ-octalactone. The OAVs of five compounds, including ethyl 2-methylbutanoate, methyl 2-methylbutanoate, ethyl 3-methylthiopropanoate, ethyl hexanoate and decanal, were the characteristic aroma compounds for the Tainong No. 6 cultivar.

Influence of Processing Conditions

The aroma profile of many commercial pineapple juices of both the single-strength juices, as well as the products made from concentrates, showed much less detectable volatile compounds in comparison with those found in juices made from fresh-cut fruits [47]. The juices made from concentrates showed some compounds produced during thermal treatment (pasteurization), such as furfural, 3-hydroxy-[2*H*]-pyran-2-one, pantolactone, 2,5-dimethyl-4-hydroxy-3(2*H*)-furanone and 5-hydroxy-methylfurfural [47].

Although minimally processed fruit and vegetables are one of the major growing segments in food retail establishments, fresh-cut fruits are still studied, because of the difficulties in preserving their fresh-like quality over long periods.

Smooth Cayenne fruits were dried in normal and modified atmospheres by addition of 0.5% v/v ethanol to the air stream, using two different temperatures and drying rates [55]. Changes in volatile compounds during the drying process were determined using HS-SPME combined with GC-MS. Modification of the drying atmosphere with ethanol vapor promoted more intense water evaporation, and a less pronounced loss of volatiles. The volatile composition changed significantly during drying, not only by the loss of some components but also by the emergence of others. It is clearly observed that drying in modified atmosphere by addition of 0.5% v/v ethanol promoted a less marked volatile loss, and more intense water evaporation. Faster water evaporation, as well as higher retention of volatiles during drying under modified atmosphere probably occurred due to ethanol condensation on the sample surface during drying.

The effects of modified atmosphere packaging on aroma compound content and the physicochemical and antioxidant attributes of cv. Gold fresh-cut pineapples imported from Costa Rica were evaluated by Montero-Calderón et al. [57]. Experiments were assessed throughout storage at 5°C. Fresh-cut pieces were packed under LO (low oxygen, 12% O_2, 1% CO_2), AIR (20.9% O_2) and HO (high oxygen, 38% O_2) headspace atmospheres. Methyl butanoate, methyl 2-methylbutanoate, and methyl hexanoate were the most abundant volatiles regardless of the packaging atmosphere and days of storage; whereas most odor active volatiles were methyl and ethyl 2-methylbutanoate, 2,5-dimethyl-4-methoxy-3(2*H*)-furanone and ethyl hexanoate. Storage life of fresh-cut pineapple was limited to 14 days due to volatile compound losses and fermentation processes.

The effect of UV-induced stress on the aroma compounds in fresh-cut pineapple was compared with that of storage at 4°C for 24 h [53]. Eighteen constituents were isolated using HS-SPME in fresh-cut pineapple and analyzed by GC-MS. Methyl-2-methylbutanoate, methyl hexanoate, methyl 5-hexenoate, ethyl hexanoate and ethyl 5-hexenoate were the major aroma compounds. Storage with these conditions and exposure of cut fruit to UV radiation for 15 min caused a considerable decrease in esters concentration and increase in the relative amount of copaene. This sesquiterpene hydrocarbon, when added to crushed cantaloupe melon (0.1 mg/g), inhibited microbial growth in the fruit over a period of 24 h at 20°C. *Cis*- and *trans*-β-ocimene were present in the fruit but their production was not photo-induced by UV irradiation. However, β-ocimene was a potent antimicrobial agent that killed microorganisms when added to the crushed fruit and stored at 20°C for 24 h. Results indicate that sesquiterpene phytoalexins could contribute to the defense mechanism in wounded pineapple tissue.

The effect of initial headspace (IH) oxygen level on the shelf-life of fresh-cut pineapple was evaluated by Zhang et al. [52]. The results showed that although the IH oxygen level had a minor effect on the growth of *Candida argentea*, *Candida sake* and *Meyerozyma caribbica* on pineapple agar, the amount of volatile organic metabolites produced by these yeasts was generally smaller the lower the IH oxygen level. The only exception was the production of ethyl acetate by *C. argentea*, which was higher at low IH oxygen levels. In triangle tests with trained judges, fruit cubes packaged in an IH of 5% oxygen were determined to be significantly different to those packaged in 21% oxygen from the 5th day of storage. Preference was shown for pineapple cubes packaged in an IH of 5% oxygen. Results suggest that packaging in an IH oxygen level of 5% could be used to extend the shelf-life of fresh-cut pineapple.

The dried fruits market is continuously increasing because of consumer preference for healthy food. However, undesirable changes can and do occur in the quality attributes of heat-sensitive materials during drying (darkening, loss of rehydration ability, shrinkage and aroma loss). Modification of the atmosphere during drying is commonly used in fruit storage for keeping the aroma, color, freshness of food and for avoiding contamination by dangerous microorganisms. Braga et al. [54] investigates the aroma composition of fresh pineapple from Brazil and samples dried under normal and modified atmosphere conditions by adding 0.5% v/v of ethanol to the drying air. Fruit rings were dried under normal and modified atmospheres in a lab-scale tunnel dryer. The atmosphere was modified by addition of 0.5% ethanol v/v to the air stream using two different temperatures and air velocities. The modified atmosphere promoted faster water evaporation and better retention of the volatile compounds upon drying. Furfural and phenylacetaldehyde were found in all dried samples but in none of the fresh samples. Higher volatiles retention was verified for the experiments performed under modified atmosphere. A large amount of ethanol in the samples dried under modified atmosphere indicated the possibility of ethanol condensation on the sample surface. The authors concluded that further studies should be carried out to get a better basic understanding of the role of ethanol in water evaporation as well as in volatiles retention.

The potential use of a commercial electronic nose in monitoring freshness of minimally processed fruit (packaged pineapple rings) during storage was examined by Torri et al. [81]. Gold Ripe pineapples from Costa Rica were taken at the beginning of their commercial life and stored at three different temperatures (4-5, 7-8, and 15-16°C) for 6-10 days. Measurements were performed by applying two analytical approaches using an electronic nose: a discontinuous method being a series of analyses on samples taken at various stages of storage and a continuous method, in which the headspace around the fruit was automatically monitored by the electronic nose probe during preservation of rings in a storage cell. The results obtained by the discontinuous approach showed that the electronic nose could discriminate between several samples and to monitor the changes in volatile compounds related to quality decay. Results revealed that fruit freshness was maintained for about five days at 5.3°C, three days at 8.6°C and one day at 15.8°C. Moreover, from the time–temperature tolerance chart, a Q_{10} value of 4.48 was derived. These data were confirmed applying the continuous method: fruit freshness was maintained for about five days at 4°C, two days at 7.6°C and one day at 16°C. The authors claimed that an in-line application of the continuous electronic nose technique could be an interesting future development.

Changes of aroma compounds of Tainong 17 fruits, cultivated in Guandong province (China), were examined during postharvest storage [60]. The volatile components were isolated by HS-SPME and analyzed by GC-MS during postharvest storage (at the 1st, 6th and 9th days after harvest) at 25°C. A total of 18 compounds were identified in which esters predominated, and methyl butanoate, methyl hexanoate, as well as methyl 3-methylthiopropanoate, were detected in both pulp and core of the fruit. In postharvest storage, total ester content increased from 65.47 to 81.18% in the pulp at the beginning; later, a decrease took place in the core. At the first day, the amount of methyl hexanoate was the highest in both pulp and core, followed by methyl butanoate in the pulp. At the sixth day, methyl butanoate was dominant in the pulp, followed by methyl hexanoate and methyl octanoate, while methyl hexanoate was the highest in the core followed by methyl octanoate and methyl butanoate. At the ninth day, methyl hexanoate and methyl 2-methylbutanoate were the major compounds in the pulp, while methyl butanoate and methyl hexanoate were the major ones in the pulp. Methyl butanoate and methyl hexanoate were predominant in the core. At the last day, ester content reached the maximum in the pulp (81.18%) and the minimum in the core (47.13%).

Changes in the aroma compounds of pineapple during freezing and thawing were compared against fresh samples to determine the effect of freezing on pineapple flavor [64]. An HS-SPME-GC-MS analysis showed that the cv. Smooth Cayenne cultivated in Thailand had 19 constituents, with four classes of compounds including 14 esters, two hydrocarbons, two sulfur-containing compounds and one lactone. The key characteristic volatile compounds of the fresh fruit were methyl hexanoate, ethyl hexanoate, ethyl 3-methylthiopropanoate and 1-(*E,Z*)-3,5-undecatriene. Freeze-thaw cycles were related with the loss of some aroma compounds, particularly the esters, which are the most characteristic compounds of fresh pineapples. The freezing and thawing process caused damage to fruit tissues due to ice recrystallization and dehydration which lead to the reduction of volatile compounds.

Osmotic dehydration of fruits by immersion is a process that consist of placing a fruit in direct contact with a solution having a suitable solute at high concentration, so that the difference in water activity values will result in two major flows in opposite directions: water moves from the fruit tissues into the solution and solute moves from the solution into the fruit tissues. This process can be used as a pretreatment to improve the sensory quality of fruits which are to be subsequently preserved by other methods, such as refrigeration, freezing and conventional drying [82]. The influence of processing variables on quality characteristics of fruits processed by this technology has been less studied than its influence on mass transfer. Some technological parameters, such as temperature, process time and both composition and concentration of the osmotic solution, have been widely studied and identified as factors affecting mass transfer during osmotic dehydration. Pino et al. [32] examined the selection of technological parameters in osmotic dehydration of pineapple by means of multivariate techniques applied to aroma compounds. These were isolated by simultaneous distillation-extraction and analyzed by GC-MS. Eleven major components were selected for multivariate analysis. Assessment of volatile losses allowed discrimination between samples obtained by different osmotic treatments. Treatments at 30°C for 1 h at atmospheric pressure or under vacuum at 30-40°C for 1-3 h with pulsed vacuum (1 cycle) produced the lowest losses of volatile compounds. Apparently, pulsed vacuum allows higher volatiles retention than the other pressure modes. It also allows increases in both process temperature and time without an important increase in losses.

Important losses of agricultural products occur across the entire productive chain from harvest to consumption. An example is pineapple residues, mainly the rind, which is generated during mechanical peeling still containing a large amount of pulp, normally disposed of with other types of residues. Since residue disposal generates considerable amounts of organic material, better economic and technical methods to dispose of these residues should be explored. Keeping in mind this idea, Barretto et al. [65] studied ways to extract and identify aroma compounds from pineapple residues generated during concentrated juice processing. Distillates of fruit residues were obtained using simple hydrodistillation and hydrodistillation by passing nitrogen gas. The compounds present in the distillates were isolated by the HS-SPME technique. Pineapple residues contain mostly esters (35%), followed by ketones (26%), alcohols (18%), aldehydes (9%), acids (3%) and other compounds (9%). Odor-active compounds were mainly identified in the distillate obtained using hydrodistillation by passing nitrogen gas; they were decanal, ethyl octanoate, acetic acid, hexan-1-ol, and lactones such as γ-hexalactone, γ-octalactone, δ-octalactone, γ-decalactone and γ-dodecalactone. The authors suggested that the use of an inert gas and lower temperatures helped maintain larger amounts of aroma compounds. These data indicate that pineapple processing residues contain important volatile compounds that can be extracted and used as aroma-enhancing products and have high potential to produce value-added natural essences.

PRODUCTION OF PINEAPPLE WINE

In the tropics, there is an abundant supply of exotic fruits to be consumed fresh or used by the food industry. However, large quantities are still wasted during peak harvest periods due to rapid post-harvest deterioration. Selection and utilization of these fruits for wine-making offer an alternative means for utilizing the fruits [83-85]. Only few studies have been conducted on the use of pineapple for wine-making, which provide a basis for further exploring into fruit wine fermentation to meet the consumers' demand for new types of fruit wine [58, 86-88].

Pino and Queris [58] developed an analytical procedure based on HS-SPME combined with GC-MS for the extraction and quantification of aroma compounds from pineapple wine. Different sample preparation variants (considering fiber type, sodium chloride addition, extraction time and temperature) were evaluated to optimize the method. In the preferred method, 8 mL of pineapple wine were placed in a 15-mL headspace vial with addition of 1 g of NaCl; a 100 µm PDMS fiber was used for extraction at 30°C for 30 min with continuous stirring. The volatiles have shown linearity in the range of concentrations studied with higher regression coefficients and reproducibility, expressed as relative standard deviation, ranged from 4.2% (2-phenylethyl acetate) to 7.1% (ethyl benzoate). The values obtained for detection and quantification limits were low enough to permit the determination of volatiles in pineapple wine. Using the optimized procedure, 18 compounds were identified, including 13 esters, four alcohols and one acid. Ethyl octanoate, ethyl acetate, 3-methylbutan-1-ol and ethyl decanoate were the major constituents. A tentative study to estimate the contribution of the identified compounds to the aroma of the wine, based on their OAVs, showed that the potentially most important compounds in pineapple wine included ethyl octanoate, ethyl acetate and ethyl 2-methylpropanoate.

The production and quality evaluation of pineapple fruit wine was examined by Ningli et al. [88]. The juice was inoculated with 5% (v/v) active yeast and held at 20°C for 7 days. Total sugar and pH decreased while alcoholic strength increased with increasing duration of fermentation. The fermented fruit wine contains 2.29 g/L total acid, 10.2% (v/v) alcohol, 5.4 °Brix soluble solids and pH 3.52. A total of 68 volatile components were found in pineapple wine, including 34 esters and 13 alcohols. The ester material accounted for 52.25% of the main aroma components. The quality and sensory evaluation results indicated that pineapple fruit wine belongs to a kind of low alcohol wine, a characteristic anticipating wide public acceptance.

Roda et al. [87] reported the processing of pineapple peel and core into quality wines by combining physical and enzymatic treatments of waste and alcoholic fermentation of the pineapple must with three strains of *Saccharomyces cerevisiae* (TT, AW, EM2) at three different temperatures. The main parameters of the alcoholic fermentation were monitored; fixed and volatile compounds of wines were analyzed by HPLC and GC-MS. The highest ethanol concentration values, i.e., more than 7 and 8% v/v, were reached in 96 h when fermentation was carried out at 20 and 25°C with AW and TT strains, respectively. The fermentation variant at 15°C with EM2 achieved the highest ethanol concentration (7.6% v/v) after 120 h and achieved higher levels of citric and malic acids. As ethanol concentration rises, an immediate decrease in simple sugars was observed; glucose fell faster than fructose and reached concentrations below 1 g/L after 120 h of fermentation. Volatile compounds were determined by HS-SPME. Significantly different aroma profiles were determined in the wines by changing temperature and strain of *S. cerevisiae*: fermentation with the AW strain at 20°C produced the highest concentration of acetate and ethyl esters, increasing the fruity character of pineapple wine, whilst varietal aroma was enhanced by fermentations at both 25 and 15°C with the TT and EM2 strains, respectively. No variations were found in pH and acidity during fermentation, while acetic acid levels were very low in all pineapple wine samples.

CONCLUSION

Pineapple is consumed worldwide and is cultivated in many tropical countries. Ripe fruits are best for eating fresh, but many processed products can be prepared, including wine. Among its notable nutritional characteristics, it is an excellent source of carbohydrates, vitamins C and A, minerals, dietary fiber, etc. Over 400 volatile constituents have been identified in fresh and processed fruit. Thus, pineapple flavor consists of a huge variety of volatile compounds. Among them, a few odorants were considered odor-active compounds. However, factors such as ripening season, fruit maturity stage, varieties, and processing conditions can directly affect the flavor profile. Moreover, information about changes in the odor-active compounds due to these factors is scarce. Studies on this subject are still very limited, so more efforts should be made not only to determine the influence of these factors on the odor-active compounds, but also to study the changes occurring during processing and storage, as well as in pre- and postharvest practices.

REFERENCES

[1] Coppens d'Eeckenbrugge, G.; Sanewski, G. M.; Smith, M. K.; Duval, M. F.; Leal, F. Ananas. In *Wild Crop Relatives: Genomic and Breeding Resources, Tropical and Subtropical Fruits*; Kole C., Ed.; Springer-Verlag: Berlin Heidelberg, 2011, pp. 21−41.

[2] Rohrbach, K. G.; Leal, F.; Coppens d'Eeckenbrugge, G. History, distribution and world production. In *The Pineapple: Botany, Production and Uses*; Bartholomew, D., Pauli, R. E., Rohrbach, K. G., Eds.; CABI: Wallingford, UK, 2003; pp. 1–12.

[3] FAO (2016). FAO database. URL http://faostat.fao.org/. Accessed 27 April 2018.

[4] Po, L. O.; Po, E. C. Tropical Fruit I: Banana, Mango, and Pineapple. In *Handbook of fruits and fruit processing*; Sinha, N. K.; Sidhu, J. S., Barta, J., Wu, J. S. B., Cano, M. P., Eds.; John Wiley & Sons, Ltd.: Oxford, UK, 2012; pp. 565–590.

[5] Morton, J. Pineapple. In *Fruits of warm climates*; Morton, J., Ed.; Julia F. Morton: Miami, FL; 1987; pp. 18–28.

[6] USDA Food Composition Databases. URL http://www.nal.usda.gov/fnic/foodcomp/search/. Accessed 20 April 2018.

[7] Steingass, C. B.; Carle, R.; Schmarr, H. G. Ripening-dependent metabolic changes in the volatiles of pineapple (*Ananas comosus* (L.) Merr.) fruit: I. Characterization of pineapple aroma compounds by comprehensive two-dimensional gas chromatography-mass spectrometry. *Anal. Bioanal. Chem.*, 2015, 407, 2591–2608.

[8] Wardencki, W.; Michulec, M.; Curyło, J. A review of theoretical and practical aspects of solid-phase microextraction in food analysis. *Int. J. Food Sci. Technol.*, 2004, 39, 703–717.

[9] Reineccius, G. A. Choosing the correct analytical technique in aroma analysis. In *Flavour in Food*; Voilley, A.; Etiévant, P., Ed.; Woodhead Publishing: Cambridge, 2006; pp. 81–97.

[10] Le Quéré, J. L. Advanced analytical methodology. In *Handbook of Fruit and Vegetable Flavors*; Hui, Y. H., Ed.; John Wiley & Sons, Inc.: Hoboken, NJ, 2010; pp. 177–194.

[11] Pessoa, F. L. P.; Mendes, M. F.; Queiroz, E. M.; Vieia de Melo, S. A. B. Extraction and distillation. In *Handbook of Fruit and Vegetable Flavors*; Hui, Y. H., Ed.; John Wiley & Sons, Inc.: Hoboken, NJ, 2010; pp. 195–210.

[12] Coelho, G. L. V.; Mendes, M. F.; Pessoa F. L. P. Flavor extraction: Headspace, SDE, or SFE. In *Handbook of Fruit and Vegetable Flavors*; Hui, Y. H., Ed.; John Wiley & Sons, Inc.: Hoboken, NJ, 2010; pp. 211–228.

[13] Jeleń, H. H.; Majcher, M.; Dziadas, M. Microextraction techniques in the analysis of food flavor compounds: A review. *Anal. Chim. Acta*, 2012, 738, 13–26.

[14] Li, J.; Wang, Y. B.; Li, K. Y.; Cao, Y. Q.; Wu, S.; Wu, L. Advances in different configurations of solid-phase microextraction and their applications in food and environmental analysis. *Trends Anal. Chem.*, 2015, 72, 141–152.

[15] Souza-Silva, É. A.; Gionfriddo, E.; Pawliszyn, J. A critical review of the state of the art of solid-phase microextraction of complex matrices II. Food analysis. *Trends Anal. Chem.*, 2015, 71, 236–248.

[16] Flath, R. Pineapple. In *Tropical and Subtropical Fruits, Composition, Properties and Uses*; Nagy, S., Shaw, P., Eds.; AVI Publishing Co.: Westport, 1980, pp. 157–183.

[17] Engel, K.-H.; Heidlas, J.; Tressl, R. The flavour of tropical fruits (banana, melon, pineapple). In *Food Flavours. Part C. Flavour of Fruits*; Morton I. D.; MacLeod, A. J., Eds.; Elsevier: Amsterdam, 1990, pp. 195–219.

[18] Berger, R. Pineapple. In *Volatile Compounds in Foods and Beverages*; Maarse, H., Ed.; Marcel Dekker Inc.: New York, NY, 1991, pp. 283–304.

[19] Nijssen, L. M.; Ingen-Visscher, C. A.; Van Donders, J. J. H. *Volatile Compounds in Food.* Online Database, Version 9.2; 2007; TNO: Zeist.

[20] Montero-Calderón, M.; Rojas-Graü, A.; Martín-Belloso, O. Pineapple (*Ananas comosus* [L.] Merril) flavor. In *Handbook of Fruit*

and Vegetable Flavors; Hui, Y. H., Ed.; John Wiley & Sons, Inc.: Hoboken, NJ., 2010; pp. 391–414.

[21] Reineccius, G. *Flavor Chemistry and Technology*; Taylor & Francis Group: Boca Raton, 2006; pp. 33−66.

[22] Haagen-Smit, A. J.; Kirchner, J. G.; Prater; A. N.; Deasy, C. L. Chemical studies of pineapple (*Ananas sativus* Lindl.). I. The volatile flavor and odor constituents of pineapple. *J. Amer. Chem. Soc.*, 1945, 67, 1646−1650.

[23] Haagen-Smit, A. J.; Kirchner, J. G.; Deasy, C. L.; Prater, A. N. Chemical studies of pineapple (*Ananas sativus* Lindl.). II. Isolation and identification of sulfur-containing esters in pineapple. *J. Amer. Chem. Soc.*, 1945, 67, 1651−1652.

[24] Gawler, J. H. Constituent of canned Malayan pineapple juice. *J. Sci. Food Agric.*, 1962, 13, 57−61.

[25] Connell, D. W. Volatile flavoring constituents of the pineapple. *Australian J. Chem.*, 1964, 17, 130−140.

[26] Rodin, J. O.; Himel, C. M.; Silverstein, R. M.; Leeper, R. W.; Gortner, W. A. Volatile flavor and aroma components of pineapple. l. Isolation and tentative identification of 2,5-dimethyl-4-hydroxy-3(2*H*)-furanone. *J. Food Sci.*, 1965, 30, 280−285.

[27] Flath, R. A.; Forrey, R. R. Volatile components of Smooth Cayenne pineapple. *J. Agric. Food Chem.*, 1970, 18, 306–309.

[28] Takeoka, G. R.; Buttery, R. G.; Teranishi, R.; Flath, R. A.; Güntert, M. Identification of additional pineapple volatiles. *J. Agric. Food Chem.*, 1991, 39, 1848−1851.

[29] Umano, K.; Hagi, Y.; Nakahara, K.; Shoji, A.; Shibamoto, T. Volatile constituents of green and ripened pineapple (*Ananas comosus* [L.] Merr.). *J. Agric. Food Chem.*, 1992, 40, 599−603.

[30] Ohta, H.; Kinjo, S.; Osajima, Y. Glass capillary gas chromatographic analysis of volatile components of canned Philippine pineapple juice. *J. Chromatogr.*, 1987, 409, 409−412.

[31] Takeoka, G.; Buttery, R. G.; Flath, R. A.; Teranishi, R.; Wheeler, E. L.; Wieczorek, R. L.; Giintert, M. Volatile constituents in pineapple (*Ananas comosus* [L.] Merr.) In *Flavor Chemistry: Trends and*

Developments. ACS Ser. 388; Teranishi, R.; Buttery, R. G.; Shahidi, F., Eds.; American Chemical Society: Washington DC, 1989; pp. 223−237.

[32] Pino, J.; Castro, D.; Fito, P.; Barat, J.; López, F. Selection of technological parameters in the osmotic dehydration of pineapple. *J. Food Qual.*, 1999, 22(6), 653−661.

[33] Pino, J. Odour-active compounds in pineapple (*Ananas comosus* [L.] Merrill cv. Red Spanish). *Int. J. Food Sci. Technol.*, 2013, 48, 564–570.

[34] Silverstein, R. M.; Rodin, J. O.; Himel, C. M.; Leeper, R. W. Volatile flavor and aroma components of pineapple. II. Isolation and identification of chavicol and *gamma*-caprolactone. *J. Food Sci.*, 1965, 30, 668−672.

[35] Rodin, J. O.; Coulson, D. M.; Silverstein, R. M.; Leeper, R. W. Volatile flavor and aroma components of pineapple. III. The sulfur-containing components. *J. Food Sci.*, 1966, 31, 721−725.

[36] Creveling, R. K.; Silverstein, R. M.; Jennings, W. G. Volatile components of pineapple. *J. Food Sci.*, 1968, 33, 284–287.

[37] Näf-Muller, R.; Willhalm, B. Uber die flüchtigen Anteile der Ananas. [Concerning the volatile fraction of pineapple]. *Helv. Chim. Acta*, 1971, 34, 1880−1890.

[38] Pickenhagen, W.; Velluz, A.; Passerat, J. P.; Ohloff, G. Estimation of 2,5-dimethyl-4-hydroxy-3(2*H*)-furanone (furaneol) in cultivated and wild strawberries, pineapples and mangoes. *J. Sci. Food Agric.*, 1981, 32, 1132−1134.

[39] Nitz, S.; Drawert F. Vorkommen von Allylhexanoat in Ananasfrüchten. [Occurrence of allyl hexanoate in pineapple fruits]. *Chem. Mikrobiol. Technol. Lebensm.*, 1982, 7, 148−152.

[40] Berger, R. G.; Drawert, F.; Nitz, S. Sesquiterpene hydrocarbons in pineapple fruit. *J. Agric. Food Chem.*, 31, 1237−1239, 1983.

[41] Berger, R. G.; Drawert, F.; Kollmannsberger, H.; Nitz, S.; Schraufstetter, B. Novel volatiles in pineapple fruit and their sensory properties. *J. Agric. Food Chem.*, 1985, 33, 233–235.

[42] Wu, P.; Kuo, M. C.; Hartman, T. G.; Rosen, R. T.; Ho, C. T. Free and glycosidically bound aroma compounds in pineapple (*Ananas comosus* L. Merr.). *J. Agric. Food Chem.*, 1991, 39, 170−172.

[43] Teai, T.; Claude-Lafontaine, A.; Schippa, C.; Cozzolino, F. Volatile compounds in fresh pulp of pineapple (*Ananas comosus* [L.] Merr.) from French Polynesia. *J. Essent. Oil Res.*, 2001, 13(5), 314−318.

[44] Preston, C.; Richling, E.; Elss, S.; Appel, M.; Heckel, F.; Hartlieb, A.; Schreier, P. On-line gas chromatography combustion/pyrolysis isotope ratio mass spectrometry (HRGC-C/P-IRMS) of pineapple (*Ananas comosus* L. Merr.) volatiles. *J. Agric. Food Chem.*, 2003, 51, 8027-8031.

[45] Brat, P.; Thi Hoang, L. N.; Soler, A.; Reynes, M.; Brillouet, J. M. Physicochemical characterization of a new pineapple hybrid (FLHORAN41 cv.). *J. Agric. Food Chem.*, 2004, 52, 6170–6177.

[46] Sinuco, D. C.; Morales, A. L.; Duque, C. Componentes volátiles libres y glicosídicamente enlazados del aroma de la piña (*Ananas comosus* L.) var. Perolera. [Free and glycosidically bound volatile components from pineapple (*Ananas comosus* L.) var. Perolera]. *Rev. Colomb. Quím.*, 2004, 33(1), 47−56.

[47] Elss, S.; Preston, C.; Hertzig, C.; Heckel, F.; Richling, E.; Schreier, P. Aroma profiles of pineapple fruit (*Ananas comosus* [L.] Merr.) and pineapple products. *LWT*, 2005, 38, 263–274.

[48] Tokitomo, Y.; Steinhaus, M.; Büttner, A.; Schieberle, P. Odor-active constituents in fresh pineapple (*Ananas comosus* [L.] Merr.) by quantitative and sensory evaluation. *Biosci. Biotechnol. Biochem.*, 2005, 69, 1323−1330.

[49] Howard, G. E.; Hoffman, A. A study of the volatile flavoring constituents of canned Malaysian pineapple. *J. Sci. Food Agric.*, 1967, 18, 106−110.

[50] Spanier, A. M.; Flores, M.; James, C.; Lasater, J.; Lloyd, S.; Miller, J. A. Fresh-cut pineapple (*Ananas* sp.) flavor. Effect of storage. In *Food Flavors: Formation, Analysis and Packaging Influences*; Contis, E. T.; Ho, C.-T.; Mussinan, C. J.; Patliment, T. H.; Shahidi,

F.; Spanier, A. M., Eds.; Elsevier Science B.V.: Amsterdam, 1998; pp. 331–343.

[51] Akioka, T.; Umano, K. Volatile components and characteristic odorants in headspace aroma obtained by vacuum extraction of Philippine pineapple (*Ananas comosus* [L.] Merr. In *Food Flavors. Chemistry, Sensory Evaluation, and Biological Activity. ACS Symp. Ser. 988*; Tamura, H.; Ebeler, S. E.; Kubota, K.; Takeoka, G. R., Eds.; American Chemical Society: Washington DC; 2008, pp. 57–67.

[52] Zhang, B.-Y.; Samapundo, S.; Rademaker, M.; Noseda, B.; Denon, Q.; de Baenst, I.; Sürengil, G.; De Baets, B.; Devlieghere, F. Effect of initial headspace oxygen level on growth and volatile metabolite production by the specific spoilage microorganisms of fresh-cut pineapple. *LWT - Food Sci. Technol.*, 2014, 55, 224–231.

[53] Lamikanra, O.; Richard, O. A. Storage and ultraviolet-induced tissue stress effects on fresh-cut pineapple. *J. Sci. Food Agric.*, 2004, 84, 1812–1816.

[54] Braga, A. M. P.; Pedroso, M. P.; Augusto, F.; Silva, M. A. Volatiles identification in pineapple submitted to drying in an ethanolic atmosphere. *Drying Technol.*, 2009, 27(2), 248–257.

[55] Braga, A. M. P.; Silva, M. A.; Pedroso, M. P.; Augusto, F.; Barata, L. E. S. Volatile composition changes of pineapple during drying in modified and controlled atmosphere. *Int. J. Food Eng.*, 2010, 6(1), Article 12. DOI: 10.2202/1556-3758.1835.

[56] Montero-Calderón, M.; Rojas-Graü, M. A.; Martín-Belloso, O. Aroma profile and volatiles odor activity along Gold cultivar pineapple flesh. *J. Food Sci.*, 2010, 75, S506–S512.

[57] Montero-Calderón, M.; Rojas-Graü, M. A.; Aguiló-Aguayo, I.; Soliva-Fortuny, R.; Martín-Belloso, O. Influence of modified atmosphere packaging on volatile compounds and physicochemical and antioxidant attributes of fresh-cut pineapple (*Ananas comosus*). *J. Agric. Food Chem.*, 2010, 58, 5042–5049.

[58] Pino, J.; Queris, O. Analysis of volatile compounds of pineapple wine using solid-phase microextraction techniques. *Food Chem.*, 2010, 122, 1241–1246.

[59] Liu, C.; Liu, Y. G..; Yi, G.; Li, W.; Zhang, G. A comparison of aroma components of pineapple fruits ripened in different seasons. *Afr. J. Agric. Res.*, 2011, 6(7), 1771–1778.

[60] Wei, C. B.; Liu, S. H.; Liu, Y. G.; Liu, X. P.; Zang, X. P.; Lv, L. L.; Sun, G. M. Changes and distribution of aroma volatile compounds from pineapple fruit during postharvest storage. *Acta Hortic.*, 2011, 902, 431–436.

[61] Wei, C. B.; Liu, S. H.; Liu, Y. G.; Lv, L. L.; Yang W. X.; Sun, G. M. Characteristic aroma compounds from different pineapple parts. *Molecules*, 2011, 16, 5104–5112.

[62] Pedroso, M. P.; Ferreira, E. C.; Hantao, L. W.; Bogusz Jr., S.; Augusto, F. Identification of volatiles from pineapple (*Ananas comosus* L.) pulp by comprehensive two-dimensional gas chromatography and gas chromatography/mass spectrometry. *J. Sep. Sci.*, 2011, 34, 1547–1554.

[63] Zheng, L. Y.; Sun, G. M.; Liu, Y. G.; Lv, L. L.; Yang, W.-X.; Zhao, W.-F.; Wei,C.-B. Aroma volatile compounds from two fresh pineapple varieties in China. *Int. J. Mol. Sci.*, 2012, 13, 7383–7392.

[64] Kaewtathip, T.; Charoenrein, S. Changes in volatile aroma compounds of pineapple (*Ananas comosus*) during freezing and thawing. *Int. J. Food Sci. Technol.*, 2012, 47, 985–990.

[65] Barretto, L. C. O.; Moreira, J. J. S.; dos Santos, J. A. B.; Narain, N.; dos Santos, R. A. R. Characterization and extraction of volatile compounds from pineapple (*Ananas comosus* L. Merril) processing residues. *Food Sci. Technol, Campinas*, 2013, 33(4), 638–645.

[66] Wei, C. B.; Ding, X. D.; Liu, Y. G.; Zhao, W. F.; Sun, G. M. Application of solid-phase microextraction for the analysis of aroma compounds from pineapple fruit. *Adv. Mat. Res.*,2014, 988, 397–406.

[67] Steingass, C. B.; Grauwet, T.; Carle, R. Influence of harvest maturity and fruit logistics on pineapple (*Ananas comosus* [L.] Merr.) volatiles assessed by headspace solid phase microextraction and gas chromatography–mass spectrometry (HS-SPME-GC/MS). *Food Chem.*, 2014, 150, 382–391.

[68] Steingass, C. B.; Langen, J.; Carle, R.; Schmarr, H. G. Authentication of pineapple (*Ananas comosus* [L.] Merr.) fruit maturity stages by quantitative analysis of γ- and δ-lactones using headspace solid-phase microextraction and chirospecific gas chromatography–selected ion monitoring mass spectrometry (HS-SPME–GC–SIM-MS). *Food Chem.*, 2015, 168, 496–503.

[69] Steingass, C. B.; Jutzi, M.; Müller, J.; Carle, R.; Schmarr, H. G. Ripening-dependent metabolic changes in the volatiles of pineapple (*Ananas comosus* (L.) Merr.) fruit: II. Multivariate statistical profiling of pineapple aroma compounds based on comprehensive two-dimensional gas chromatography-mass spectrometry. *Anal. Bioanal. Chem.*, 2015, 407, 2609–2624.

[70] Chaintreau, A. Simultaneous distillation-extraction: from birth to maturity. Review. *Flavour Fragr. J.*, 2001, 16, 136–148.

[71] Engel, W.; Bahr, W.; Schieberle, P. Solvent assisted flavor evaporation – a new and versatile technique for the careful and direct isolation of aroma compounds from complex food matrices. *Eur. Food Res. Technol.*, 1999, 209, 237–241.

[72] IOFI Working Group on Methods of Analysis. Guidelines for solid-phase micro-extraction (SPME) of volatile flavour compounds for gas-chromatographic analysis, from the Working Group on Methods of Analysis of the International Organization of the Flavor Industry (IOFI). *Flavour Fragr. J.*, 2010, 25, 404–406.

[73] Tressl, R.; Engel, K. H.; Albrecht, W.; Bille-Abdullah, H. Analysis of chiral aroma components in trace amounts. In *Characterization and Measurement of Flavor Compounds. ACS Symp. Ser. 289*; Bills, D.D.; Mussinan, C., Eds.; American Chemical Society: Washington, D.C., 1985; pp. 43−60.

[74] Tressl, R.; Heidlas, J.; Albrecht, W.; Engel, K.-H. Biogenesis of chiral hydroxyacid esters. In *Bioflavour '87*; Schreier, P., Ed.; Walter de Gruyter, Berlin, New York, 1988, p. 221−236.

[75] Engel, K. H.; Heidlas, J.; Albrecht, W.; Tressl, R. Biosynthesis of chiral flavor and aroma compounds in plants and microorganisms. *ACS Symp. Ser.*, 1989, No. 388, 8−22.

[76] Schieberle, P. New developments in methods for analysis of volatile flavor compounds and their precursors. In *Characterization of Food: Emerging Methods*; Gaonkar, A. G., Ed.; Elsevier Science B. V.: Amsterdam, 1995; pp. 403−431.

[77] Pitet, A. O.; Rittersbacher, P.; Muralidhara, R. Flavor properties of compounds related to maltol and isomaltol. *J. Agric. Food Chem.*, 1970, 18, 929−933.

[78] Re, L.; Maurer, B.; Ohloff, G. Ein einfacher Zugang zu 4-Hydroxy-2,5-dimethyl-3(2*H*)-furanone (Furaneol) einem Aromastandteil von Ananas und Erdbere. [An easy access to 4-hydroxy-2,5-dimethyl-3(2*H*)-furanone (furaneol), a flavoring component of pineapple and strawberry]. *Helv. Chim. Acta*, 1973, 56, 1882−1884.

[79] Burdock, G. A. *Fenaroli's Handbook of Flavor Ingredients*; Taylor and Francis Group, LLC: Boca Raton, FL., 2010.

[80] Crouzet, J.; Chassagne, D. Glycosidically bound volatiles in plants. In *Naturally Occurring Glycosides*; Ikan, R., Ed.; Wiley: New York, 1999; pp. 225−274.

[81] Torri, L.; Sinelli, N.; Limbo, S. Shelf life evaluation of fresh-cut pineapple by using an electronic nose. *Postharvest Biol. Technol.*, 2010, 56, 239–245.

[82] Ahmed, I.; Qazi, I. M.; Jamal, S. Developments in osmotic dehydration technique for the preservation of fruits and vegetables. *Innov. Food Sci. Emerg. Technol.*, 2016, 34, 29–43.

[83] Duarte, W. F.; Dias, D. R.; Oliveira, J. M.; Teixeira, J. A.; Almeida e Silva, J. B.; Schwan, R. F. Characterization of different fruit wines made from cacao, cupuassu, gabiroba, jaboticaba and umbu. *LWT - Food Sci. Technol.*, 2010, 43, 1564−1572.

[84] Pino, J.; Queris, O. Analysis of volatile compounds of mango wine. *Food Chem.*, 2010, 125, 1141−1146.

[85] Pino, J.; Queris, O. Characterisation of odour-active compounds in papaya (*Carica papaya* L.) wine. *Int. J. Food Sci. Technol.*, 2012, 47, 262–268.

[86] Idise, O. E. Studies of wine produced from pineapple (*Ananas comosus*). *Int. J. Biotechnol. Mol. Biol. Res.*, 2012, 3(1), 1–7.

[87] Roda, A.; De Faveri, D. M.; Dordoni, R.; Valero-Cases, E.; Nuncio-Jauregui, N.; Carbonell-Barrachina, A. A.; Frutos-Fernández, M. J.; Lambri, M. Pineapple wines obtained from saccharification of its waste with three strains of *Saccharomyces cerevisiae*. *J. Food Process. Preserv.*, 2016, 40, doi: 10.1111/jfpp.13111.

[88] Ningli, Q.; Lina, M.; Liuji, L.; Xiao, G.; Jianzhi, Y. Production and quality evaluation of pineapple fruit wine. *IOP Conf. Series: Earth and Environmental Science 100* (2017) 012028 doi: 10.1088/1755-1315/100/1/012028.

In: The Pineapple
Editor: Lydia Hampton

ISBN: 978-1-53614-594-6

Chapter 2

IRRADIATED PINEAPPLE AND FRUIT QUALITY

Apiradee Uthairatanakij[1,*], Pongphen Jitareerat[1], Sukanya Aiamla-or[2] and Peter A. Follett[3]

[1]School of Bioresources and Technology, King Mongkut's University of Technology Thonburi, Bangkok, Thailand

[2]Ratchaburi Learning Park, King Mongkut's University of Technology Thonburi, Ratchaburi, Thailand

[3]USDA-ARS, US Pacific Basin Agricultural Research Center, Hilo, Hawaii, US

ABSTRACT

Internal browning or blackheart is an important physiological disorder of fresh pineapple fruit occurring during low temperature storage and a limiting factor during export. Phytosanitary treatments including heat, cold, or ionizing radiation can be used to control quarantine pests for market access. Irradiation is a viable alternative for insect disinfestation and can prolong the storage life of fresh fruits and

* Corresponding Author Email: apiradee.uth@kmutt.ac.th.

vegetables. The effects of gamma irradiation on the quality of fresh produces may vary with species and cultivar, maturity stage, harvest season and radiation dose. In pineapple fruit, irradiation can increase the severity of internal browning symptoms beyond the effects of cold storage alone. Pre and postharvest treatments and conditions (maturity stage and harvesting season) are important factors that may be manipulated to reduce the severity of internal browning in irradiated cold-stored pineapples and postharvest losses. Certain fruit coatings also may help reduce internal browning symptoms and need further investigation.

Keywords: internal browning, ionizing irradiation, postharvest treatment, phytosanitary, quality

1. Introduction

Pineapple (*Ananas comosus* Merr.), belongs to the *Bromeliaceae* family, is an economically important exotic tropical fruit available throughout the year. The fibrous yellow seedless flesh has a sweet flavor and is the source of important vitamins and minerals and of the enzyme bromelain, which may alleviate joint pain and reduce inflammation [1]. Pineapple is a non-climacteric fruit and needs to be harvested at higher maturity to ensure optimal edible quality [2]. Pineapple fruit maturity and quality are mainly judged by sugar to acid ratios, skin color and aroma compounds [3]. Fruit is perishable and susceptible to physiological disorders such as internal browning resulting in a short shelf life. Internal browning (IB) is a major physiological disorder of pineapple fruit associated with the polyphenol oxidase activity that often occurs after low temperature storage and most pineapple varieties are susceptible [4, 5, 6, 7]. Water soaked patches are initially found in the fruit flesh close to the core, and these patches subsequently become brown, and eventually browning extends from the core outward into the fruit flesh. This symptom develops faster after the fruit is transferred from cold storage to warmer physiological temperatures (15-30°C) [8, 9]. Some varieties such as 'Gold' or 'MD-2' are less prone to IB. Methods to completely or partially alleviate

IB include heat treatment, controlled atmospheres, polyethylene bagging, salicylic acid application 1-MCP and waxing [10, 11, 12, 13, 14].

Coating or waxing has been applied to maintain the quality and reduce postharvest diseases of fresh fruits such as apple [15], citrus [16] and pineapple [17]. Coatings create modified atmospheres by changing the concentrations of CO_2 and O_2 inside the fruit, and also form a barrier to reduce water loss [18, 3]. However, the efficacy of coatings depends on the materials and methods [15, 19]. Coating treatments delayed pineapple fruit ripening and maintained fruit quality [20], and reduced IB of pineapple under low temperature stresses via maintenance of cell integrity [14, 17] reported that coatings reinforced the antioxidant system and improved fruit quality mainly through the reduction of organic acid content. In this chapter, we investigate the effects of cultivar, irradiation dose, fruit maturity, harvest season and coating treatments on the quality of gamma irradiated pineapple fruit.

The source of ionizing radiation (irradiation) treatment can be high-energy electrons or X-rays (machine generated), or gamma rays (from cobalt-60 or cesium-137) [21]. Gamma radiation has been considered as an alternative to the use of chemicals for inhibiting the growth of pathogenic microorganisms, for delaying fruit ripening and extending shelf life, and for insect disinfestation [22, 23]. For insect disinfestation, a minimum irradiation dose of 0.15 kGy was approved by the IPPC for tephritid fruit flies [24], and a minimum dose of 0.4 kGy was approved by the United States for all insects except pupae and adults of Lepidoptera [25]. Food irradiation has been widely endorsed and approved as a sanitary and phytosanitary method for over 60 foods and food products by major food and health organizations such as the World Health Organization (WHO), the Food and Agriculture Organization (FAO), the International Atomic Energy Agency (IAEA), the Centers for Disease Control and Prevention (CDC) and Codex Alimentarius Commission (CAC) [26]. For phytosanitary uses, the publication in 2003 of the International Standard for Phytosanitary Measures (ISPM) no. 18 "Guidelines for the use of irradiation as a phytosanitary measure" marked a significant step forward in overcoming regulatory barriers for the adoption of radiation technology as a means to facilitate international trade in horticultural products [27]. The implementation of this standard in practice has resulted in exports of irradiated mango, longan, mangosteen, rambutan, dragon fruit, and guava from India, Mexico, Pakistan, Thailand and Vietnam to the United States, and the export of mangoes from Australia to New Zealand [23]. Moreover, irradiation can be used as *"hurdle technology"* in order to maintain the fruit quality and also extend shelf-life [28]. The number of countries using phytosanitary irradiation to expand agricultural trade is rapidly increasing, as are the number of different fruits and vegetables being exported. The export of pineapple from Thailand to the United States has been approved by applying a generic minimum dose of 0.4 kGy [29].

2. Irradiation Affecting Fruit Quality

In recent years, the effect of using γ-irradiation as a phytosanitary treatment on postharvest quality has been studied in many types of produce. The minimum dose for pests has been already approved in the IPPC standard No. 28, however, the maximum acceptable dose during commercial treatment will depend on fruit tolerance, which is influenced by cultivar, maturity stage, pre-harvest conditions, time of harvesting, postharvest conditions, and interactions among these factors [30]. In some cases irradiation may enhance the quality of fresh produce. For example, in a study conducted by [31], irradiation at 0.4 and 1.0 kGy increased the content of organic acids in Kishu mandarins. [32] reported that higher total soluble solids, total and reducing sugars, and ascorbic acid content, and minimum acidity, were found in 'Kesar' mango fruit irradiated with 0.4 kGy and stored at 12°C compared to non-irradiated fruits stored at ambient condition at ripening stage [33] showed that blueberry fruit treated with 1.0 kGy irradiation maintained fruit firmness and colour, and reduced the loss of nutrients. Irradiation up to 2.5 kGy in jujube fruit caused a significant increase in the total monomeric anthocyanin and the total phenolic content [34]. Strawberries treated with 1.5 kGy were also reportedly higher in anthocyanin levels [35, 36] reported that gamma irradiation at 1.0 kGy was able to slower respiration and ethylene production, and improved lipoxygenase (LOX) activity. In contrast, [37] reported that irradiation at 0.5, 1.0 and 2.0 kGy did not affect respiratory rate, ethylene production, flesh firmness, anthocyanin content or color index of the raspberries. Application of carboxymethyl cellulose (CMC) coating at 1.0% (w/v) followed by irradiation at 1.5 kGy resulted in chlorophyll retention of plums, as resulting in delay fruit ripening [38, 39] also reported that 'Tommy Atkins' mangos irradiated with 1.0 kGy remained in a less advanced stage of ripening (stage 3) which reduced its overall quality.

Radiation treatment is often reported to improve quality, but in some studies, there was not a significant effect on physico-chemical changes in the fruit after applying irradiation. The tolerance and sensitivity of fresh produce depends on irradiation dose, maturity stage and type of commodity. According to [40] mature green mangoes cv. Nam Dokmai and Chok Anan at the 70% and 90% maturity were subjected to ionizing irradiation in the range of 0.4 to 0.6 kGy. Ripened irradiated "Nam Dokmai" mangoes appeared firmer compared to untreated fruit, but irradiation had no effect on skin or flesh color and soluble solids content of mangoes of both cultivars harvested at both maturity stages [41] demonstrated that strawberries irradiated up to 2.0 kGy presented the same hedonic mean on acceptance tests during 12 days of storage compared to control fruit, but the control strawberries presented better results for the ascorbic acid content, firmness and weight loss. Likewise, vitamin C was significantly decreased in 1.0 kGy irradiated Korea citrus fruit at day 20 of storage [42]. Irradiated raspberries cv. Autumn Bliss at 0.5-2.0 kGy showed only a minor reduction in ascorbic acid content [37]. In 'Mosambi' sweet orange, radiation dose up to 1.5 kGy reduced acidity and vitamin C content [43]. Radiation treatment reduced ascorbic acid and β-carotene content in guavas var. Media China [31, 44] reported that irradiation at 0.4 kGy and 1.0 kGy caused immediate reductions in pulp firmness, vitamin E, individual sugars and carotenoids in 'Kishu' mandarins. Irradiation at dose 0.8 kGy of freshly packed cherry tomatoes (*Solanum lycopersicum* var. Cerasiforme), caused a decrease in firmness compared with non-irradiated fruit [45, 46] reported that irradiated strawberries cv. Amado and Marquee at 0.4 kGy were on average of 23% softer than control fruit. In contrast, the TSS content was higher in irradiated ‘Nagpur’ mandarin (*Citrus reticulata* Blanco), ‘Mosambi’ sweet orange (*Citrus sinensis* Osbeck) and ‘Kagzi’acid lime (*Citrus aurantifolia* Swingle) fruit when exposed with dose up to 1.5 kGy [43].

Irradiation can be used not only for delayed ripening, but also for microbial safety and preventing decay. Gamma irradiation at 1.0 kGy was reported to reduce the disease and decay incidence in ‘Autumn Bliss’ raspberries stored at 0°C and 90% RH [37]. Irradiation doses at 1.0 kGy to 2.5 kGy were considered to be a feasible method for reducing the fruit rot rate in blueberry fruit [33, 47] also reported that blueberry fruit irradiated by 2.5 kGy had reduced decay after storage at 5°C for 35 days compared to unirradiated controls [48] found that cherries (Misri and Double) treated with doses of 1.2 or 1.5 kGy had no decay during ambient storage for 9 days. The combination of 1.0% (w/v) Carboxymethyl cellulose (CMC) and 1.5 kGy irradiation was beneficial in delaying the decay of plum during post-refrigerated ambient storage at 25 ± 2°C and RH 70% [38]. On the other hand, irradiation at 0.4 kGy and 1 kGy enhanced browning of the calyx end and fungal infection in ‘Kishu’ mandarins [31].

The postharvest quality of some fresh commodities may remain unaffected at various doses of irradiation. For example, brix, organic acids, and sensory properties of Korean citrus fruits, ‘Jinjihyang’ and ‘Chunggyun’ were not affected by gamma irradiation with dose of 1.0 kGy [42]. Titratable acidity, soluble solids, color values, and ascorbic acid content were unchanged in 0.4 kGy of gamma irradiated ‘Amado’ and ‘Marquee’ strawberries [46], blueberry [49], and grapes [50]. Similar results have been reported by [51], who suggested 1.0 and 1.5 kGy might be used as consumers acceptable doses for shelf life extension, minimum weight loss and decay, without affecting the chemical quality of strawberry cv. Corona. [52] reported that irradiation at ≤1 kGy did not affect quality (overall fruit quality, colour, firmness, shrivel, weight loss, TSS, TA levels, TSS/TA ratio and juice pH), or the nutritional or proximate content (contents of ash, carbohydrate, dietary fibre, energy, moisture, protein, sodium, potassium, total sugars, fructose, ascorbic acid, anthocyanin, citric and malic acids) of treated blueberry fruit. There was no significant difference between irradiated 'Ataulfo' mango (0.15-0.3 kGy) in terms of weight loss, external and internal color, pH, soluble solids, titratable acidity and firmness, and sensorial quality [53]. Phytosanitary irradiation doses (0.15-0.4 kGy) did not negatively impact blueberry and sweet cherry quality [54], did not significantly affect the postharvest quality of different blueberries varieties and slightly improved shelf life [30], and showed no difference in liking for irradiated 'Amado' and 'Marquee' strawberries compared to the control fruit [46, 49] also showed that irradiation doses of ≤1000 Gy did not significantly affect blueberry or raspberry fruit quality (overall fruit quality, colour, firmness, weight loss, TSS, TA levels or TSS/TA ratio), or the nutritional or proximate content. In addition, irradiation at 0.25-0.75 kGy did not affect colour and percent sugar of tomato [55].

For the consumer, the quality of fresh produce is primarily evaluated subjectively by appearance, color, flavor and texture. Deterioration of these properties caused by irradiation may affect the purchase or repurchase of the product. Brushed 'Sufaid Chaunsa' mangoes irradiated at 0.7 kGy had significantly higher lenticel development and poor colour than controls [56, 57] reported dose-dependent effects in 'Lane Late' navel oranges in terms of visual damage (increased external damage and pitting with increased dose) and increased weight loss. Similarly, in 'Mosambi' sweet orange, radiation treatments at 0.25, 0.5, 1.0 and 1.5 kGy caused peel disorder in the form of brown sunken areas after 90 days, but not in'Nagpur' mandarin [43, 58] noted that acetic acid fumigation increased the negative effects on quality of irradiated fruit due to increased cracks in the fruit peel of 'Daw' longan fruit during storage at 4°C. Irradiation doses of 0.25 kGy and 0.75 kGy negatively affected skin quality and pulp of exposed Tahitian lime fruits, and doses >0.1 kGy caused skin yellowing in the normally green fruit [59]. 'Sufaid Chaunsa' mangoes subjected to irradiation doses of 0.7 and 1.0 kGy had significantly less peel colour development compared to lower doses of 0.25 and 0.4 kGy [56]. There was no adverse effect of color appearance, odor and flavor in 'Ataulfo' mango irradiated with gamma radiation at 0.15 kGy and 0.3 kGy [53].

3. Irradiated Pineapple Fruit

The US Food and Drug Administration (FDA) currently restricts the maximum irradiation level for fresh fruits and vegetables to 1.0 kGy, with only two exceptions of fresh lettuce and spinach that can be irradiated up to 4.5 kGy [60, 61]. Phytosanitary irradiation for pineapple must not only control quarantine pests to gain market access, but also maintain postharvest quality including its overall physical, biochemical, and organoleptic qualities, and ideally extend storage life as well. The major obstacle for use of irradiation for phytosanitary or other purposes in pineapple is the variation in response to maturity stage, harvest season, and cultivar. 'Trad Si Thong' and 'Pattavia' are commonly grown pineapple cultivars in Thailand. Phenolic content in irradiated 'Trad Si Thong' pineapple fruits at 0.4 kGy were higher than the control fruit. However, the dose of gamma irradiation significantly affected the total soluble solids (TSS)/titratable acids (TA) ratio and the antioxidant activity [62, 63] reported that gamma irradiation at 0.40 kGy did not affect the ratio of TSS/TA, antioxidant content, or ascorbic acid content in 'Pattavia' pineapple fruit. In addition, radiation at dose 0.5 kGy induced browning and softening tissue of pineapple fruit [64]. [65] reported that minimally processed 'Smooth Cayenne' pineapple irradiated with gamma irradiation at 0 (control), 1.0 and 2.0 kGy showed no significant differences in texture, color, acidity and vitamin C content, but significantly increasing pH and TSS during storage at 8°C for 10 days. Consideration of various factors influencing the postharvest quality of irradiated pineapples follows.

3.1. Effects of Cultivar, Maturity Stage and Season Factors on the Quality of Irradiated Pineapple

The quality and characteristics of pineapple fruit are dependent on several natural factors including the cultivar, maturity stage and harvest season. For example, under the same growing conditions, the quality of pineapple cv. Trad Sri Thong is different from cv. Pattavia, due to genetic factors. Pulp color of 'Trad Sri Thong' is a more pronounced yellow color than 'Pattavia' as indicated by b* and Hue angle values, (Table 1). 'Trad Sri Thong' pineapple also has higher titratable acid, ascorbic acid, polyphenol content and antioxidant capacity (% DPPH) than 'Pattavia,' whereas 'Pattavia' has higher sweetness (higher total soluble solids, TSS) than 'Trad Sri Thong.' 'Trad Sri Thong' is more sensitive than 'Pattavia' to cold temperature storage at 13°C for 14 days, exhibiting a higher incidence of internal browning symptoms (also called "chilling injury"). The physico-chemical and antioxidant capacity of pineapple fruit may be changed after irradiation treatment depending on cultivar. Color of pulp (lightness, yellow and hue angle), TSS, and DPPH of pineapple cv. Pattavia pulp decreased when the fruits were irradiated with gamma irradiation at 0.3-0.6 kGy, whereas TA, vitamin C, and polyphenol content increased. On the other hand, gamma irradiation increased the severity of internal browning and also affected the decrease of TA and polyphenol content of 'Trad Sri Thong' during cold storage (Table 1).

Another factors affecting the quality of irradiated pineapple is fruit maturity. [66] demonstrated the effect of gamma irradiation on the quality of pineapple 'Trad Sri Thong' fruit when harvested at the different maturity stages: early stage (full green), middle stage (slightly color break) and lately stage (1/4-1/2 gold) (Table 2). Gamma irradiation at 0.3-0.6 kGy did not influence pulp color of any maturity stage, but did affect internal browning. Early maturity stage showed higher internal browning than the middle and late stages. Thus, the maturity stage of pineapple fruit which may be most suitable for gamma treatment is the late stage. Maturity stage also affected on the physicochemical characters of gamma irradiated pineapple. Late stage gamma-irradiated pineapple had lower TA and higher vitamin C, polyphenol content and DPPH than middle and early stage fruit, whereas TSS concentrations did not different between maturity stages. Respiration rate and ethylene production of all maturity stages was not significant different, but gradually increased throughout storage.

Several studies have shown that harvesting season may be an important factor on fruit quality [67] reported that pulp color (lightness, b* and hue angle values) and TSS of 'Pattavia' pineapple harvested in summer season was not significant different with pineapple harvested in winter and rainy season. However, pineapple harvested in winter season had higher TSS and vitamin C than the other seasons. Harvesting season affected the severity of internal browning and DPPH activity. The pineapple harvested in the rainy season had the lowest internal browning and highest DPPH activity during storage at 13°C for 14 days (Table 3).

Table 1. Cultivar factors affecting on the physiological and biochemical changes of gamma irradiated pineapple fruits (cv. Pattavia and Trad Sri Thong). All fruits were kept in dark condition at 13°C for 14 days

Cultivars	Treatment	Days after GI	Pulp color			IB	TA	TSS	AsA	PP	DPPH
			*L**	*b**	Hue°						
Pattavia	Non GI	0	81.50	24.02	101.43	0.0	1.24	15.15	30.11	10.64	25.10
		7	78.86	23.01	100.69	0.0	1.47	16.39	28.04	13.24	28.67
		14	75.75	22.36	100.61	0.3	1.93	15.36	24.11	13.75	32.82
	GI (0.3-0.6 kGy)	0	77.61	14.64	98.53	0.0	2.07	12.20	33.76	12.63	22.74
		7	76.97	17.54	99.48	0.0	2.20	13.03	35.24	14.72	26.33
		14	74.23	19.45	100.76	0.0	2.37	12.13	44.56	13.26	26.41
Trad Sri Thong	Non GI	0	77.32	35.28	96.70	0.0	2.58	14.50	85.33	34.62	54.46
		7	75.74	34.19	92.41	0.0	2.71	14.17	69.55	19.23	72.55
		14	70.26	33.45	95.91	1.8	2.41	13.62	64.63	29.17	66.32
	GI (0.3-0.6 kGy)	0	76.66	28.76	94.24	0.0	2.33	14.50	85.33	19.58	54.17
		7	78.51	33.20	94.48	0.0	2.11	14.95	95.53	20.85	59.09
		14	76.26	34.75	93.43	2.8	1.93	14.63	50.77	21.44	58.97

GI; Gamma irradiated fruit, IB; Internal browning (score), TA; Total acidity content (%), TSS; Total soluble solid content (°brix), AsA; Ascorbic acid content (mg/100gFW), PP; Polyphenol content (mg/100gFW), DPPH; 1,1-diphenyl-2-picrylhydrazyl) radical scavenging activity (%)

This may be result of growing conditions such as degree of temperature and relative humidity. Although the lowest degree of internal browning was observed in pineapple harvested in the rainy season, when fruits harvested in rainy season were treated with gamma irradiation, the degree of internal browning was higher than the irradiated fruits harvested in summer and winter. Our findings highlight the importance of harvesting season on the quality of irradiated pineapples. Gamma irradiation also caused on the decrease of pulp lightness in pineapples that were harvested in summer and winter but not in the rainy season. Gamma irradiation had no effect on TA, TSS, polyphenol content and DPPH activity in any harvested seasons but harvesting season did the effect on DPPH activity.

Table 2. Harvesting maturity factors affecting on the physiological and biochemical changes of gamma irradiated pineapple fruits cv. Trad Sri Thong. All fruits were kept in dark condition at 13°C for 14 days

Maturity stages	Days after GI	Pulp color L*	b*	Hue°	IB	TA	TSS	AsA	PP	DPPH	RS	C_2H_4
Early	0	76.95	24.79	95.46	0.0c	2.24a	17.27a	35.46c	16.01b	56.47a	2.58c	2.58c
	7	77.94	28.27	96.89	1.5b	2.00a	14.23b	56.83b	16.37b	45.57ab	5.20c	7.38a
	14	76.60	31.46	95.06	3.6a	1.55b	13.40b	33.21c	17.58b	30.66b	9.08a	4.47ab
Middle	0	76.66	27.49	94.24	0.0c	2.07a	14.50b	85.33a	17.59b	53.87a	2.66c	2.38c
	7	78.90	31.75	95.13	0.0c	1.93a	14.17b	64.86	18.67ab	46.99ab	5.79c	5.34ab
	14	76.92	33.81	93.75	2.9a	1.58b	13.62b	30.57c	20.96a	53.77a	8.61ab	4.22b
Lately	0	78.18	34.72	93.52	0.0c	1.94ab	14.40b	96.64a	16.18b	53.50a	3.15c	2.57c
	7	76.18	37.11	93.39	0.0c	1.89ab	14.33b	62.39b	19.43ab	58.30a	7.62b	6.14a
	14	73.91	37.98	92.26	2.4ab	1.56b	13.77b	42.54bc	26.21a	55.57a	10.72a	5.05ab
F-test		ns	ns	ns	*	*	*	*	*	*	*	*

GI; Gamma irradiation, IB; Internal browning (score), TA; Total acidity content (%), TSS; Total soluble solid content (°brix), AsA; Ascorbic acid content (mg/100gFW), PP; Polyphenol content (mg/100gFW), DPPH; 1,1-diphenyl-2-picrylhydrazyl) radical scavenging activity (%), RS; respiration rate (mgCO_2/kg.hr), C_2H_4; Ethylene production of fruit (ul/kg.hr)

3.2. Effect of Irradiation Dose on the Quality of Pineapple

Irradiation of commodities can be used for controlling insect infestations, inhibition of seed germination, postharvest disease control, and extension of the shelf life. However, dose tolerance information is often not available or limited for particular crops. Various morphological, structural and functional changes of the fruit may be dose dependent. Higher irradiation doses may create abnormal symptoms or tissue injury [64] revealed that the maximum dose tolerance for pineapple fruit cv. Queen was about 0.25 kGy. When applying the required 0.4 kGy minimum dose to export fruit to the United States, the maximum and minimum dose will range between 0.4-1.0 kGy [66] reported that the high irradiation dose

Table 3. Harvesting season factors affecting on the physiological and biochemical changes of gamma irradiated pineapple fruits cv. Pattavia. All fruit samples were kept in dark condition at 13°C for 14 days

Harvesting season	Treat-ment	Days after GI	Pulp color			IB	TA	TSS	AsA	PP	DPPH
			L*	b*	Hue°						
Summer	Non GI	0	72.99a	17.82b	97.76	0.0c	1.14b	14.50	33.91b	11.95ab	22.65b
		7	75.32a	21.45ab	100.07	0.0c	1.43b	14.61	31.23b	13.97a	26.87b
		14	73.75a	25.11a	99.43	0.8ab	1.87b	13.90	28.81b	12.02a	38.90a
	GI (0.3-0.6kGy)	0	71.78a	20.20ab	99.58	0.0c	1.31b	14.23	28.34b	13.67a	20.27b
		7	63.57b	21.24ab	100.12	0.1c	1.00b	14.45	28.65b	14.10a	22.90b
		14	62.37b	19.91ab	100.88	1.3a	1.45b	14.74	44.95a	13.65a	31.29a
Winter	Non GI	0	69.99ab	26.33a	96.39	0.0c	2.80a	14.00	32.36b	8.06b	24.13b
		7	75.03a	24.13a	95.66	0.5b	2.96a	13.78	45.00a	13.36a	31.16a
		14	75.99a	19.82ab	99.15	0.7ab	3.63a	14.04	40.92a	12.91a	25.21b
	GI (0.3-0.6 kGy)	0	76.99a	20.28ab	97.75	0.0c	2.82a	14.08	39.17ab	11.58a	25.20b
		7	76.72a	23.37a	96.63	0.8b	3.40a	14.32	41.82a	15.34a	29.76a
		14	69.28ab	18.40b	99.48	2.9a	3.28a	13.48	44.17a	12.86a	21.52b
Rainy	Non GI	0	74.50a	19.91ab	100.05	0.0c	1.19b	14.46	30.11b	10.63ab	25.69b
		7	73.62a	21.11ab	100.77	0.0c	1.09b	13.86	26.05c	13.24a	30.83a
		14	75.89a	22.97a	101.38	0.3b	1.45b	14.40	25.89c	13.78a	33.00a
	GI (0.3-0.6kGy)	0	74.50a	19.91ab	100.05	0.0c	1.19b	14.46	39.11a	10.63ab	25.70b
		7	73.85a	20.37ab	100.71	0.0c	1.42b	14.25	28.04ab	13.24a	28.67ab
		14	70.69a	22.09a	101.44	1.1ab	1.20b	14.21	25.76c	13.78a	32.82a
F-test			*	*	ns	*	*	ns	*	*	*

GI; Gamma irradiation, IB; Internal browning (score), TA; Total acidity content (%), TSS; Total soluble solid content (°brix), AsA; Ascorbic acid content (mg/100gFW), PP; Polyphenol content (mg/100gFW), DPPH; 1,1-diphenyl-2-picrylhydrazyl) radical scavenging activity (%)

of 1.0 kGy did not cause changes in pulp color of pineapple when compared with the dose of 0.5 kGy treated fruits and non-treated fruits. But at the higher dose of 1.0 kGy, the severity of internal browning was significantly higher than lower dose at the lower dose of 0.5 kGy. However, there were no significant differences in the changes of TA, TSS, vitamin C, polyphenol content, DPPH activity, ethylene production, respiration rate, and enzymatic browning activities (polyphenol oxidase and peroxidase) between gamma irradiation dose of 0.5 and 1.0 kGy.

Physiological and biochemical changes of pineapple fruits are dependent on the duration of storage time (Table 4).

Table 4. Effect of different doses of gamma irradiation on physiological and biochemical changes of pineapple fruit cv. Trad Sri Thong during storage in dark condition at 13°C

Treat-ment	Days after GI	Pulp color			IB	TA	TSS	AsA	PP	DPPH	RS	C_2H_4	BE	
		L*	b*	Hue°									PPO	POD
Non GI	0	76.66	30.03	94.24	0.0b	2.58	14.50	85.33b	21.57a	54.46c	2.41c	2.18c	0.43b	0.56b
	7	78.10	30.94	94.77	0.0b	2.71	14.17	98.23ab	25.15a	76.24a	4.42b	8.57b	0.92b	0.70b
	14	75.60	33.66	93.85	2.83a	2.41	13.62	64.63c	13.34b	66.32b	8.15a	5.33b	3.40a	0.82ab
0.5 kGy	0	76.66	30.03	94.24	0.0b	2.58	14.50	85.33b	21.57a	54.46c	2.41c	2.18c	0.43b	0.56b
	7	78.11	34.65	93.82	0.0b	2.28	15.73	126.20a	23.02a	71.19a	3.44b	11.47a	0.70b	0.68b
	14	75.60	35.68	93.10	2.67a	2.28	15.63	70.96bc	21.92a	64.17b	7.03a	4.53b	2.89a	1.04a
1.0 kGy	0	76.66	30.03	94.24	0.0b	2.58	14.50	85.33b	21.57a	54.46c	2.41c	2.18c	0.43b	0.56b
	7	79.85	32.11	94.66	0.0b	2.56	14.17	122.37a	27.33a	73.71a	3.84b	11.56a	0.53b	0.32b
	14	75.43	35.03	93.67	3.67a	2.26	14.67	63.96c	18.98a	64.91b	7.10a	4.57b	1.85a	0.61b
F-test		ns	ns	ns	*	ns	ns	*	*	*	*	*	*	*

GI; Gamma irradiation, IB; Internal browning (score), TA; Total acidity content (%), TSS; Total soluble solid content (°brix), AsA; Ascorbic acid content (mg/100gFW), PP; Polyphenol content (mg/100gFW), DPPH; 1,1-diphenyl-2-picrylhydrazyl) radical scavenging activity (%), RS; respiration rate ($mgCO_2$/kg.hr), C_2H_4; Ethylene production of fruit (ul/kg.hr), BE; Browning enzyme activities, PPO; Polyphenol peroxidase (unit/mg protein), POD; Peroxidase (unit/mg protein).

3.3. Reducing Postharvest Losses of Gamma Irradiated Pineapple Fruit

Internal browning is a major barrier to long-term storage of pineapple. Exporting fruit to the United States for example by surface shipment may require three weeks in cold storage, leading to high postharvest loss. Several studies on pre- and post-harvested treatments have been investigated to reduce this internal browning of pineapple, which may be especially important since irradiation may increase internal browning under certain conditions. Coatings and waxing fruit are known to reduce respiration rate resulting in quality maintenance of many types of fresh commodities [67] showed that coating pineapples with sucrose fatty acid ester (SFE) and starfish 7055 retarded the respiration rate. An application of 10% SFE also significantly retarded internal browning in 'Pattavia' pineapple, but resulted in decreased DPPH activity. Coating pineapple with SFE and Starfesh 7055 at the concentrations of 5-10% did not have any significant effects on the changes of pulp color, TA, TSS, and polyphenol contents (Table 5). The polyphenol composition of tropical fruits highly varies with the species, the cultivar and the tissue and can be changed by storage conditions and processing [68].

Table 5. Effect of coating substances on physiological and biochemical changes of gamma irradiated pineapple cv. Pattavia. All fruit samples were treated with gamma ray at 0.3-0.6 kGy and kept in dark condition at 13°C for 14 days

Treat-ment	Days after GI	Pulp color			IB	TA	TSS	AsA	PP	DPPH	RS
		L*	b*	Hue°							
GI	0	74.50	19.91	100.05	0.0c	1.86	14.46	30.11a	10.63	25.70b	12.02b
	7	73.62	20.37	100.71	0.0c	2.23	14.25	28.04a	13.24	28.67a	29.57a
	14	75.85	22.09	101.44	1.1a	1.89	14.21	24.18ab	13.77	32.82a	17.49ab
SF	0	74.50	19.91	100.05	0.0c	1.86	14.46	30.11a	10.63	25.70b	12.02b
7055	7	75.65	21.83	100.66	0.0c	2.09	15.64	22.50b	13.93	30.75a	27.52a
5%	14	72.86	20.20	101.61	0.9a	2.03	14.65	24.56ab	13.89	33.30a	20.10a
SF	0	74.50	19.91	100.05	0.0c	1.86	14.46	30.11a	10.63	25.70b	12.02b
7055	7	74.49	21.33	101.07	0.0c	1.96	15.17	23.48b	12.42	29.34a	23.12a
10%	14	68.50	21.80	101.11	0.8a	2.00	13.63	24.53ab	14.35	31.14a	13.98b
SFE	0	74.50	19.91	100.05	0.0c	1.86	14.46	30.11a	10.63	25.70b	12.02b
5%	7	74.49	20.17	100.99	0.0c	1.66	14.78	25.89a	11.07	27.60ab	21.83a
	14	68.50	21.38	100.72	0.8a	1.60	16.00	22.37b	14.93	27.95ab	10.24b
SFE	0	74.50	19.91	100.05	0.0c	1.86	14.46	30.11a	10.63	25.70b	12.02b
10%	7	75.03	20.64	100.55	0.0c	1.69	15.36	26.60a	10.31	28.69b	20.60a
	14	72.35	21.95	100.49	0.2b	1.76	14.50	26.04a	13.79	23.70b	11.08b
F-test		ns	ns	ns	*	ns	ns	*	ns	*	*

SF; Starfesh, SFE; Sucrose fatty ester, GI; Gamma irradiation, IB; Internal browning (score), TA; Total acidity content (%), TSS; Total soluble solid content (°brix), AsA; Ascorbic acid content (mg/100gFW), PP; Polyphenol content (mg/100gFW), DPPH; 1,1-diphenyl-2-picrylhydrazyl) radical scavenging activity (%), RS; respiration rate (mgCO_2/kg.hr)

The coating agents allow delaying enzymatic browning because they produce a modified atmosphere on coated fruits [69]. This result indicates that type of coating substance and its optimal concentration are required to study to fit with each crop.

CONCLUSION

Phytosanitary irradiation is a superior treatment for control of quarantine pests due to its broad effectiveness against quarantine pests, no chemical residues, possible extension of storage life, and the ability to treat product in the final packaging. A limitation in exporting pineapple under long-term cold storage is internal browning. Phytosanitary irradiation may induce physiological and chemical changes of pineapple, and increase internal browning under certain conditions. Response of pineapple to irradiation is dependent on the cultivar, maturity and harvesting season. The severity of internal browning is dose-dependent—higher doses of irradiation cause the higher internal browning. Reducing postharvest losses of irradiated pineapple fruit may be achieved by applying specific fruit coatings. The effect of cultural practices needs to investigate in the future.

REFERENCES

[1] Hossain, F. (2016). World Pineapple Production: An overview. *African Journal of Food, Agriculture, Nutrition and Development,* 16 (4): 11443-11456. http://dx.doi.org/10.18697/ajfand.76.15620.

[2] Lobo, M. G. and Yahia, E. (2017). *Biology and Postharvest Physiology of Pineapple.* John Wiley & Sons, Ltd. http://dx.doi.org/10.1002/9781118967355.ch3.

[3] Lobo, M. G. and Paull, R. E. (2017). *Handbook of Pineapple Technology: Postharvest Science, Processing and Nutrition.* John Wiley & Sons, Ltd. http://dx.doi.org/10.1002/9781118967355.

[4] Paull, R. E. and Rohrbach, K. G. (1985). Symptom Development of Chilling Injury in Pineapple Fruit. *Journal of the American Society for Horticultural Science,* 110: 100-105.

[5] Zhou, Y., Dahler, J. M. Underhill, S. J. R. and Wills, Ron B. H. (2003). Enzymes Associated with Blackheart Development in Pineapple Fruit. *Food Chemistry,* 80: 565-572.

[6] Hassan, A., Othman, Z. and Siriphanich, J. (2011). Pineapple (*Ananas comosus* L. Merr.). *Postharvest Biology and Technology of Tropical and Subtropical Fruits*, 194-218.

[7] Zhang, Q., Rao, X., Zhang, L., He, C., Yang, F. and Zhu, S. (2016). Mechanism of Internal Browning of Pineapple: The Role of Gibberellins Catabolism Gene (AcGA2ox) and GAs, *Scientific Reports*, 6 (1).

[8] Akamine, E. K., Goo, T., Steepy, T., Greidanus, T. and Iwaoka, N. (1975). Control of Endogenous Brown Spot of Fresh Pineapple in Postharvest Handling. *Journal of the American Society for Horticultural Science,* 100: 60-65.

[9] Abdullah, H. (1997). Research on Black Heart Disorder in Malaysian Pineapple. *Pineapple News* (*ISHS Pineapple Working Group Newsletter*), 111: 10-11.

[10] Rohrbach, K. G. and Paull, R. E. (1982). Incidence and Severity of Chilling Induced Internal Browning of Waxed Smooth Cayenne Pineapple. *Journal of the American Society for Horticultural Science*, 107: 453-457.

[11] Nakasone, H. Y. and Paull, R. E. (1998). *Tropical Fruits*, CAB International, Wallingford, Oxon, UK.

[12] Lu, X. H., Sun, D. Q., Li, Y. H., Shi, W. Q. and Sun, G. M. (2011). Pre- and Post-Harvest Salicylic Acid Treatments Alleviate Internal Browning and Maintain Quality of Winter Pineapple Fruit. *Scientia Horticulturae*, 130: 97-101.

[13] Selvarajah, S., Bauchot, A. D. and John, P. (2001). Internal Browning in Cold Stored Pineapples is suppressed by a Postharvest Application of 1-Methylcyclopropene. *Postharvest Biology and Technology*, 23: 167–170.

[14] Li, X., Zhu, X., Wang, H., Lin, X., Lin, H. and Chen, W. (2018). Postharvest Application of Wax Controls Pineapple Fruit Ripening and Improves Fruit Quality. *Postharvest Biology and Technology,* 136: 99-110. http://dx.doi.org/10.1016/j.postharvbio.2017.10.012.

[15] Bai, J., Hagenmaier, R. D. and Baldwin, E.A. (2003). Coating Selection for 'Delicious' and Other Apples. *Postharvest Biology and Technology,* 28: 381-390. http://dx.doi.org/10.1016/S0925-5214(02)00201-6.

[16] Fan, F., Tao, N., Jia, L. and He, X. (2014). Use of Citral Incorporated in Postharvest Wax of Citrus Fruit as a Botanical Fungicide against *Penicillium digitatum. Postharvest Biology and Technology,* 90: 52-55. http://dx.doi.org/10.1016/j.postharvbio.2013.12.005.

[17] Hu, H. G., Li, X. P., Dong, C. and Chen, W. X. (2012). Effects of Wax Treatment on the Physiology and Cellular Structure of Harvested Pineapple during Cold Storage. *Journal of Agricultural and Food Chemistry*, 60: 6613-6619.

[18] Baldwin, E. A. (2003). Coatings and Other Supplemental Treatments to Maintain Vegetable Quality. *Postharvest Physiology and Pathology of Vegetables*, 413-456.

[19] Baldwin, E., Burns, J., Kazokas, W., Brecht, J., Hagenmaier, R., Bender, R. and Pesis, E. (1999). Effect of Two Edible Coatings with Different Permeability Characteristics on Mango (*Mangifera indica* L.) Ripening during Storage. *Postharvest Biology and Technology,* 17: 215-226. http://dx.doi.org/10.1016/S0925-5214(99)00053-8.

[20] Lin, X., Li, X. and Chen, W. (2013). Effect of Wax Treatment on the Quality and Postharvest Physiology of Pineapple Fruits. *Acta Horticulturae,* 975: 519-526. http://dx.doi.org/10.17660/ActaHortic.2013.975.68.

[21] Ihsanullah, I. and Rashid, A. (2017). Current Activities in Food Irradiation as a Sanitary and Phytosanitary Treatment in the Asia and the Pacific Region and a Comparison with Advanced Countries. *Food Control,* 72: 345-359.

[22] IAEA. (1992). Use of Irradiation as a Quarantine Treatment of Food and Agricultural Commodities. In: *Proceedings of a meeting, Malaysia 1990, International Atomic Energy Agency*, Vienna.

[23] Barkai-Golan, R. and Follett, P. A. (2017). *Irradiation for Quality Improvement, Microbial Safety and Phytosanitation of Fresh Produce*. Academic Press, San Diego, CA, USA.

[24] FAO, (2009). *Phytosanitary Treatments for Regulated Pests*. ISPM No. 28. FAO, Rome.

[25] APHIS. (2006). Treatments for Fruits and Vegetables. *Fed. Regist.,* 71: 4451-4464.

[26] IFSAT. (2013). Institute of Food Science and Technology *Information Sheet Irradiation*, Surrey, UK. https://www.ifst.org/documents/misc /Irradiation2013.pdf.

[27] IPPC (International Plant Protection Convention). (2003). *International Standards for Phytosanitary Measures. Guidelines for the Use of Irradiation as a Phytosanitary Treatment.* FAO (Food Agriculture Organization), Rome.

[28] Arvanitoyannis, I. S., Stratakos, A. and Tsarouhas, P. (2009). Irradiation Applications in Vegetable and Fruit. *Critical Reviews in Food Science and Nutrition*, 49: 427-462. DOI: 10.1080/10408390 80206793658

[29] Follett, P. (2009). Generic Radiation Quarantine Treatments: The Next Steps. *Journal of Economic Entomology,* 102 (4): 1399-1406.

[30] Lires, C. M. L., Docters, A. and Horak, C. I. (2018). Evaluation of the Quality and Shelf Life of Gamma Irradiated Blueberries by Quarantine Purposes. *Radiation Physics and Chemistry,* 143: 79-84.

[31] Ornelas-Paz, J. D. J., Meza, M. B., Obenland, D., Rodríguez, K., Jain, A., Thornton, S. and Prakash, A. (2017). Effect of Phytosanitary Irradiation on the Postharvest Quality of Seedless Kishu Mandarins (*Citrus kinokuni* mukakukishu). *Food Chemistry,* 230: 712-720.

[32] Yadav, M. K., Patel, N. L. and Patel, S. R. (2013). Effect of Irradiation and Storage Temperature on Quality Parameters of Kesar Mango (*Mangifera indica* L.). *Indian Journal of Plant Physiology,* 18 (3): 313-317.

[33] Wang, C. and Meng, X. (2016). Effect of 60Co γ-Irradiation on Storage Quality and Cell Wall Ultra-Structure of Blueberry Fruit During Cold Storage. *Innovative Food Science and Emerging Technologies,* 38: 91-97.

[34] Najafabadi, N. S., Sahari, M. A., Barzegar, M. and Esfahani, Z. H. (2017). Effect of Gamma Irradiation on some Physicochemical Properties and Bioactive Compounds of Jujube (*Ziziphus jujuba* var. vulgaris) Fruit. *Radiation Physics and Chemistry,* 130: 62-68.

[35] Nassur, R. D. C. M. R., Lima, R. A. Z., Lima, L. C. O. and Chalfun, N. N. J. (2016). Gamma Radiation Doses on Strawberry Quality Maintenance. *Comunicata Scientiae,* 7 (1): 38-48.

[36] Wang, C., Gao, Y., Tao, Y., Wu, X. and Zhibo, C. (2017). Influence of γ-Irradiation on the Reactive-Oxygen Metabolism of Blueberry Fruit during Cold storage. *Innovative Food Science and Emerging Technologies,* 41: 397-403.

[37] Tezotto-Uliana, J. V., Berno, N. D., Saji, F. R. Q. and Kluge, R. A. (2013). Gamma Radiation: An Efficient Technology to Conserve the Quality of Fresh Raspberries. *Scientia Horticulturae,* 164: 348-352.

[38] Hussain, P. R., Suradkar, P. P., Wani, A. M. and Dar, M. A. (2015). Retention of Storage Quality and Post-Refrigeration Shelf-Life Extension of Plum (*Prunus domestica* L.) cv. Santa Rosa Using Combination of Carboxymethyl Cellulose (CMC) Coating and Gamma Irradiation. *Radiation Physics and Chemistry,* 107: 136-148.

[39] Cruz, J. N., Soares, C. A., Fabbri, A. D. T., Cordenunsi, B. R. and Sabato, S. F. (2012). Effect of Quarantine Treatments on the Carbohydrate and Organic Acid Content of Mangoes (cv. Tommy Atkins). *Radiation Physics and Chemistry,* 81 (8): 1059-1063.

[40] Uthairatanakij, A. Jitareerat, P. and Kanlayanarat, S. (2006). Effects of Irradiation on Quality Attributes of Two Cultivars of Mango. *Acta Horticulturae,* 712: 885-891.

[41] Jesus Filho, M. D., Scolforo, C. Z., Saraiva, S. H., Pinheiro, C. J. G., Silva, P. I. and Della Lucia, S. M. (2018). Physicochemical, Microbiological and Sensory Acceptance Alterations of Strawberries

Caused by Gamma Radiation and Storage Time. *Scientia Horticulturae,* 238: 187-194.

[42] Jo, Y., Nam, H. A., Ramakrishnan, S. R., Baek, M. E., Lim, S. B. and Kwon, J. H. (2018). Postharvest Irradiation as a Quarantine Treatment and Its Effects on the Physicochemical and Sensory Qualities of Korean Citrus Fruits. *Scientia Horticulturae,* 236: 265-271.

[43] Ladaniya, M. S., Singh, S and Wadhawan, A. K. (2003). *Response of 'Nagpur' Mandarin, 'Mosambi' Sweet Orange and 'Kagzi' Acid Lime to Gamma Radiation. Radiation Physics and Chemistry,* 67:665-675. https://doi.org/10.1016/S0969-806X(02)00480-2

[44] Rosario, R. C., Julieta, S. G., Emilia, B. G. and Valdivia-López, M. A. (2013). Irradiation Effects on the Chemical Quality of Guavas. *Advance Journal of Food Science and Technology,* 5 (2): 90-98.

[45] Guerreiro, D., Madureira, J., Silva, T., Melo, R., Santos, P. M. P., Ferreira, A., Trigo, M. J., Falcão, A. N., Margaça, F. M. A. and Cabo Verde, S. (2016). Post-Harvest Treatment of Cherry Tomatoes by Gamma Radiation: Microbial and Physicochemical Parameters Evaluation. *Innovative Food Science and Emerging Technologies,* 36: 1-9.

[46] Serapian, T. and Prakash, A. (2016). Comparative Evaluation of the Effect of Methyl Bromide Fumigation and Phytosanitary Irradiation on the Quality of Fresh Strawberries. *Scientia Horticulturae,* 201: 109-117.

[47] Wang, C., Li, X. T. and Meng, X. J. (2016). Effect of ^{60}Co-γ Irradiation on Quality and Physiology of Blueberry Storage. *Resources, Environment and Engineering Proceedings of the 2nd Technical Congress*, Taylor and Francis Group, London, pp. 417-422.

[48] Parveen, S., Hussain, P. R., Mir, M. A., Shafi, F., Darakshan, S., Mushtaq, A. and Suradkar, P. (2015). Gamma Irradiation Treatment of Cherry - Impact on Storage Quality, Decay Percentage and Post-Refrigeration Shelf-Life Extension. *Current Research in Nutrition and Food Science,* 3 (1): 54-73.

[49] Golding, J. B., Blades, B. L., Satyan, S., Jessup, A. J., Spohr, L. J., Harris, A. M., Banos, C. and Davies, J. B. (2014). Low Dose Gamma Irradiation Does Not Affect the Quality, Proximate or Nutritional Profile of 'Brigitta' Blueberry and 'Maravilla' Raspberry Fruit. *Postharvest Biology and Technology,* 96: 49-52.

[50] Kim, G. C., Rakovski, C., Caporaso, F. and Prakash, A. (2014). Low-Dose Irradiation Can be used as a Phytosanitary Treatment for Fresh Table Grapes. *Journal of Food Science,* 79 (1): S81-S91.

[51] Majeed, A., Muhammad, Z., Majid, A., Shah, A. H. and Hussain, M. (2014). Impact of Low Doses of Gamma Irradiation on Shelf Life and Chemical Quality of Strawberry (*Fragariax ananassa*) cv. 'Corona.' *Journal of Animal and Plant Sciences,* 24 (5): 1531-1536.

[52] Golding, J. B., Satyan, S., Spohr, L. J., Harris, A. M., Blades, B. L., Jessup, A. J., Banos, C. and Davies, J. B. (2015). Effects of Low Dose Gamma Irradiation on Highbush Blueberry Fruit Quality and Proximate and Nutritional Profiles. *Acta Horticulturae,* 1079: 515-520.

[53] Gómez-Simuta, Y., Hernández, E., Aceituno-Medina, M., Liedo, P., Escobar-López, A., Montoya, P., Bravo, B., Hallman, G. J., Bustos, M. E. and Toledo, J. (2017). Tolerance of Mango cv. 'Ataulfo' to Irradiation with Co-60 vs. Hydrothermal Phytosanitary Treatment. *Radiation Physics and Chemistry,* 139: 27-32.

[54] Thang, K., Au, K., Rakovski, C. and Prakash, A. (2016). Effect of Phytosanitary Irradiation and Methyl Bromide Fumigation on the Physical, Sensory, and Microbiological Quality of Blueberries and Sweet Cherries. *Journal of the science of food and agriculture,* 96 (13): 4382-4389.

[55] Akter, H. and Khan, S. A. (2012). Effect of Gamma Irradiation on the Quality (colour, firmness and total soluble solid) of Tomato (*Lycopersicon esculentum* Mill.) Stored at Different Temperature. *Asian Journal of Agricultural Research,* 6 (1): 12-20.

[56] Malik, A. U., Umar, M., Hameed, R., Amin, M., Asad, H. U., Hafeez, O. and Hofman, P. J. (2013). Phytosanitary Irradiation

Treatments in Relation to Desapping and Processing Types Affect Mango Fruit Quality. *Acta Horticulturae,* 1012: 681-692.

[57] McDonald, H., Arpaia, M. L., Caporaso, F., Obenland, D., Were, L., Rakovski, C. and Prakash, A. (2013). Effect of Gamma Irradiation Treatment at Phytosanitary Dose Levels on the Quality of 'Lane Late' Navel Oranges. *Postharvest Biology and Technology,* 86: 91-99.

[58] Boonpok, N., Uthairatanakij, A., Srilaong, V., Photchanachai, S. and Jitareerat, P. (2013). Effects of Acetic Acid Fumigation on Suppressing Postharvest Decay of Gamma Irradiated Longan Fruit. *Acta Horticulturae*, 973: 97-102.

[59] Da Silva, S. R., Bezerra, D. N. F., Bassan, M. M., Cantuarias-Avilés, T. and Arthur, V. (2016). Postharvest of Irradiated Tahiti Lime Fruits. *Revista Brasileira de Fruticultura,* 38: 9 p.

[60] Abolhassani, Y., Caporaso, F., Rakovski, C. and Prakash, A. (2013). The Effect of Gamma Irradiation as a Phytosanitary Treatment on Physicochemical and Sensory Properties of Bartlett Pears. *Journal of Food Science*, 78: 1437–1444.

[61] Shahbaz, H. M., Akram, K., Ahn, J. J. and Kwon, J. H. (2016). Worldwide Status of Fresh Fruits Irradiation and Concerns about Quality, Safety, and Consumer Acceptance. *Food Science & Nutrition*, 56(11): 1790-807.

[62] Uthairatanakij, A., Jitareerat, P., Srilaong, V. and Photchanachai, S. (2013). Effect of Gamma Irradiation Dose on Postharvest Quality and Antioxidant Activity of 'Trad Si Thong' Pineapple. *Acta Horticulturae,* 1012: 829-836.

[63] Jenjob, A., Uthairatanakij, A., Jitareerat, P., Wongs-Aree, C. and Aiamla-Or, S. (2017). Effect of Harvest Seasonal and Gamma Irradiation on the Physicochemical Changes in Pineapple Fruit cv. Pattavia during Stimulated Sea Shipment. *Food Science and Nutrition,* 5 (5): 997-1003.

[64] Damayanti, M., Sharma, G. J. and Kundu, S. C. (1992). *Gamma-Radiation Influences Postharvest Disease Incidence of Pineapple Fruits*. Hort Science, 27: 807-808.

[65] Perecin, T. N., Silva, L. C. A. S., Harder, M. N. C., Oliveira, A. C. S., Arevalo, R. and Arthur, V. (2011). Evaluation of the Effects of Gamma Radiation on Physical and Chemical Characteristics of Pineapple (*Ananas comosus* (L.) Meer) cv. Smooth Cayenne Minimally Processed. *Progress in Nuclear Energy,* 53: 1145-1147. doi: 10.1016/ j.pnucene.2011.06.015.

[66] Uthairatanakij, A., Jitareerat, P., Photchanachai, S., and Srilaong, V. (2011). *Effect of Gamma Irradiation on Physico-Chemical Changes of Pineapple*, Postharvest Technology Innovation center. Thailand, 79 pages.

[67] Jenjob, A. (2017). *Effects of Harvesting Seasons and Coating on Internal. Browning of Gamma Irradiated Smooth Cayenne Pineapple.* Master thesis, King Mongkut's University of Technology Thonburi.

[68] Rinaldo D., Mbe′guie D. M. A. and Fils-Lycaon, B. (2010). Advances on Polyphenols and their Metabolism in Sub-Tropical and Tropical Fruits. *Trends in Food Science & Technology,* 21: 599-606.

[69] Ioannou, I. and Ghoul, M. (2013). Prevention of Enzymatic Browning in Fruit and Vegetables. *European Scientific Journal,* 9: 1857-7881.

In: The Pineapple
Editor: Lydia Hampton

ISBN: 978-1-53614-594-6

Chapter 3

BIOACTIVE METABOLITES FROM PINEAPPLE AND THEIR NUTRITIONAL BENEFITS

Maria Terezinha Santos Leite Neta[1], Rajiv Gandhi Sathiyabama[2] and Narendra Narain[1,*]
[1]Laboratory of Flavor and Chromatographic Analysis,
Federal University of Sergipe, Sergipe, Brazil
[2]Department of Medicine, Postgraduate Program in Health Sciences,
Federal University of Sergipe, Sergipe, Brazil

ABSTRACT

Around the world, consumers are becoming much more aware of the potential health benefits of tropical fruits in human nutrition, as they are considered to be rich sources of bioactive metabolites along with their pleasant and strong aroma, appearance and taste. Pineapple is a native fruit of South America and highly appreciated for its exotic characteristics and presence of vital fruit nutrients. The nutritional and therapeutic values of this fruit are highly explored by biomedical researchers living in the South American region, which focus great interest in pineapple as a rich source of carotenoids, flavonoids, phenolic compounds and vitamin C with potential application on the reduced risk of several diseases, such as cancer, inflammation, cardiovascular, cataracts, diabetes, Alzheimer's disease, macular degeneration and neurodegenerative diseases. In addition, bromelain, a proteolytic enzyme which is present in pineapple possesses enormous pharmaceutical applications, including in anti-inflammatory, anti-diarrheal, anticancer, and inhibition of platelet aggregation. This chapter discusses various classes of bioactive metabolites from pineapple reported in the scientific literature and their nutritional properties. The effects of these metabolites on human health are vital, and hence it is extremely important to understand their nutraceutical potential and future applications. Thus, the chapter will also present the importance of pineapple fruit consumption in human diet due to the presence of important nutrients in this fruit.

Keywords: pineapple, phenolics, carotenoids, bromelain, health-promoting, diseases

1. INTRODUCTION

Pineapple is one of the most important tropical fruits in the world and it could be consumed either in fresh form or in the form of its products such as juices, jams, jellies and dried fruit. It is the third most important tropical fruit in the world after bananas and citrus, and it is highly perishable and seasonal (Bartolomew et al., 2006). The pineapple fruit is composed of water, carbohydrates, minerals, sugars and fibers that are good for the digestive system and help in maintaining ideal weight and balanced nutrition (Benitez et al., 2014).

The total world exports of fresh and canned pineapple in 2011 was 3.15 and 1.23 million tons, with exported values of 172.7 and 130.7 million US dollars (FAOSTAT, 2014). It is estimated that the pineapple export involves around 1.2 billion US dollars for countries in Asia, parts of Africa and Latin America (UNCTAD, 2012). Despite the existence of more than 100 varieties of pineapple fruit, only 6 to 8 of them are cultivated commercially. Among them, the main variety cultivated are the "Smooth Cayenne" and the low-acid hybrid "MD-2" (UNCTAD, 2012).

However, the processing industry is dominated by a number of higher acid clones of "Smooth Cayenne" variety (Valderrain-Rodriguez et al., 2017).

The composition of macronutrients in pineapple was explored by researchers, especially in its edible portion. It contains 81.2 to 86.2% moisture and 13-19% of the total solids while carbohydrates represent up to 85% of the total solids. Fiber makes up for 2-3% (Hossain et al., 2015). Besides these compounds, pineapple fruits are important sources of some vitamins such as A, B and mainly vitamin C, flavonoids, phenolic compounds and carotenoids (Alias & Abba, 2017). These compounds seek out oxidative damage in body cells by scavenging reactive oxygen species due to its antioxidant capacity, which can prevent several diseases, such as some types of cancers, inflammation, cardiovascular, cataracts, diabetes, Alzheimer's disease, macular degeneration and neurodegenerative diseases (Yeoh & Ali, 2017). This chapter reviews the scientific information available about the main bioactive compounds from pineapple and its nutritional properties.

2. Bioactive Compounds

Bioactive compounds are extra nutritional constituents that occur naturally in small quantities in plant products and lipid rich foods that are capable of modulating metabolic processes and consequently result in the promotion of better health. These compounds are intimately associated with their positive effects on human health, particularly on cardiovascular diseases, cancer and on the delay of the ageing process (Kris- Etherton et al., 2002; Galanakis, 2017; Valderrain-Rodriguez, 2017). Their positive effects are mainly attributed to antioxidant activity, inhibition or induction of enzymes, inhibition of receptor activities, besides induction and inhibition of gene expression (Correia, Borges, Medeiros, & Genovese, 2012). These compounds are grouped according to their chemical structure and function, which include carotenoids, flavonoids, phenolic acids, glucosinolates, dietary fiber, phytosterols, monoterpenes and some very active molecules including ascorbic acid (Ayala- Zavala et al., 2011). The bioaccessibility and bioavailability of each bioactive compound differ greatly, and the most abundant compounds in ingested fruit are not necessarily those leading to the highest concentrations of active metabolites in target tissues (Manach, Williamson, Morand, Scalbert, & Remesy, 2005). For pineapple, the main compounds reported are bioactive compounds with antioxidant activity, vitamins (ascorbic acid), phenolic compounds and carotenoids (Kongsuwan et al., 2009). Besides these compounds other constituents of pineapple can also act like bioactive compounds and show beneficial effects in the human health like bromelain, a complex of compounds including proteinases and dietary fiber (Maurer, 2001).

2.1. Vitamins

Vitamins are a group of organic compounds that are very essential for the human body because of their action such as coenzymes, hormones precursors in the genetics regulation, components in the defense system of the organism against oxidation and in some specific functions (Benassi & Mercadante, 2018). The vitamins are found in low concentrations in foods, but its amount is enough to supply the human needs and these are the compounds that must be present in the food to enable human growth, health and life (Paul & Shaha, 2004).

Vitamins are divided in two classes according to their solubility – fat soluble which include vitamins A, D, E and K and water soluble – vitamins complex B (thiamine, riboflavin, niacin, pantothenic acid, vitamin B6, biotin, B12) and vitamin C (Benassi & Mercadante, 2018). The main benefits of consumption of vitamins such as vitamin C, vitamin A, and vitamin E are in its antioxidant action and control of oxidative stress. Studies suggest that exogenous anti-oxidants such as vitamin E, vitamin C, carotenoids, and flavonoids can reduce beta-amyloid toxicity in patients with Alzheimer's disease and Parkinson's disease (Largani et al., 2018).

The presence of vitamins such as thiamine, riboflavin, ascorbic acid and β-carotene precursor of vitamin A (retinol) were reported (Silva et al.,

2008; Almeida et al., 2009). Paul & Shaha, (2004) reported contents of thiamine, riboflavin, β-carotene and ascorbic acid to be around 0.20; 0.12; 18.9 and 35 mg/100 g in fresh pineapple pulp, respectively. The high contents of ascorbic acid and β-carotene demonstrate that pineapple is an important source of vitamins which are essential for the human diet.

The content of vitamin C in pineapple varies from 35.8 to 62.11 mg/100 g of fresh fruit pulp (Ferreira et al., 2016). Table 1 assembles the data on vitamin C contents in pineapple, as reported by several researchers. The variation in the amount of vitamin C is occasioned due to different species and varieties especially when they grow in different environmental conditions (Ferreira et al., 2016).

Vitamin C is considered an important nutrient that occurs naturally and has antioxidant action when consumed in the diet (Almeida et al., 2011). It inhibits the development of serious clinical conditions such as heart disease and some types of cancers. This vitamin has high bioavailability and is quickly metabolized by the human body acting like one of the most important antioxidants in cells and as scavenger of reactive oxygen species (ROS). Thus, vitamin C can protect membranes and lipoproteins from oxidative damage (Sies & Stahl, 1995; Gardner et al., 2000).

Table 1. Vitamins A and B (mg/100g of fresh pulp) contents of different cultivars of Pineapple*

Cultivar	Vitamin A	Vitamin B_3	Vitamin B_6	Vitamin B_{12}
Comte de Paris	0.0038	0.0089	0.0069	—
Smooth Cayenne	0.0035	0.0060	0.0071	0.0066
Phuket	0.0025	0.0049	0.0066bc	—
MD-2	0.0024	0.0058	0.0143	0.0040
Tainung19	0.0025	0.0091	0.0302	0.0121
Comte de Paris mutant	0.0016	0.0090	0.0049	0.0084
Tainung11	0.0019	0.0038	0.0104	—
Golden Winter Sweet	0.0018	0.0235	0.0671	—
Shenwan	0.0025	0.0076	0.0045	0.0020
New Phuket	0.0024	0.0045	0.0078	—
Tainung17	0.0035	0.0076	0.0059	0.0120

*Table adapted from Sun et al. (2016).

Table 2. Vitamin C (mg/100 g of fresh pulp) contents in different cultivars of pineapple according to various publications

	Ferreira et al. (2016)	**September-Malaterre et al., (2016)**	**Hernadéz et al., (2006)**	**Gardner et al., (2000)**	**Kongswan et al., (2009)**	**Viana et al., (2013)**	**Sun et al., (2016)**
Imperial	62.11	-	-	4.4 ± 0.5	-	-	-
Pérola	49.79	-	-	-	-	21.43	-
IAC Fantástico	43.72	-	-	-	-	-	-
Smooth Cayenne	42.31	-	-	-	-	15.18	9.99
Gomo-de-mel	36.74	-	-	-	-	-	-
Thai Pulae	-				18.88	-	-
Nanglae	-				6.45	-	-
Vitória	35.88	28.9	-	-	-	16.17	-
Roja-española	-	-	26.6	-	-	-	-
Comte de Paris	-	-	-	-	-	-	17.78
Phuket	-	-	-	-	-	-	18.60
MD-2	-	-	-	-	-	-	22.39
Tainung19	-	-	-	-	-	-	16.91
Comte de Paris mutant	-	-	-	-	-	-	17.69
Tainung11							23.09
Golden Winter Sweet	-	-	-	-	-	-	11.18
Shenwan	-	-	-	-	-	-	21.45
New Phuket	-	-	-	-	-	-	16.80
Tainung17	-	-	-	-	-	-	14.28

-Not analyzed.

2.2. Phenolic Compounds

Phenolic compounds are constituents of many plants which are bioactive in the human body (Hassimoto and Mercadante, 2018). These are secondary metabolites of plants and have the protective function against factors such as pathogens, stress and UV-B radiation. These compounds can contribute also to the sensorial quality such as bitterness, astringency, flavor, color and oxidative stability of fruits and vegetables (Naczk; Shahidi, 2004).

Phenolic compounds are widely distributed in a variety of plants and in citrus fruits, such as lemon, orange and mandarin, as well as in other fruits such as cherry, grape, plum, pear, apple and papaya. Green pepper, broccoli, purple cabbage, onion, garlic and tomato are also important sources of these compounds which belong to a diversified group of phytochemicals derived from phenylalanine and tyrosine originated from the secondary metabolism of plants (Inbaraj et al., 2010). Phenolic compounds are essential for the growth and reproduction of plants and these are formed as a result of stress conditions such as infections, wounds, ultraviolet radiation. The simple phenol is the most common phenolic compound, because it has only one hydroxyl group directly attached to the aromatic ring (Soares, 2002).

The phenolic compounds composition in fruits and vegetables can be very complex to determine due to the great structural variety that these compounds have. Phenolic compounds can be classified in several groups such as flavonoids, anthocyanins, chalconoids, flavanones, anthocyanidins, flavones, isoflavones, pro-anthocyanidins. Some of these compounds can be found in some fruits and vegetables while others are more specific like isoflavones that are found in soya beans (Hassimoto and Mercadante, 2018).

These properties characterize an effect in health protection, not only by antioxidant activity by scavenging free radicals, but also inhibition of hydrolytic and oxidative enzymes and anti-inflammatory functions in human cells, thus conferring on phenolic compounds (Rocha, 2011; Naczk; Shahidi, 2004). The phenolic compounds composition of fruits can vary due to genetic and environmental factors, and it may be modified by oxidative reactions during fruit storage when two important processes such as the antioxidant activity and oxidative browning occur (Rotili et al., 2013).

Pineapple fruits also exhibit good levels of phenolic compounds as reported by Ferreira et al., (2016) in their study, about content of phenolic compounds in different cultivars. They found the contents of the total phenolics varying from 71.07 mg.gallic acid/100g in Smooth Cayenne cultivar to 126.95 mg.gallic acid/100 g in Imperial cultivar. In another work, Martínez et al., (2012) analyzed pineapple byproducts (mainly core and shell) and reported the content of phenolic compounds similar to that of Imperial cultivar, being around 129 mg.gallic acid/100g. Paz et al., (2015) showed the evaluation of bioactive compounds in several tropical fruits and reported the phenolic compounds content of 329 mg.gallic acid/100g of fresh pineapple pulp. Table 2 assembles the data on phenolic compounds contents in pineapple, as reported by several researchers.

Table 3. Total phenolic compounds (mg.gallic acid/100g) contents in different cultivars of pineapple according to various researchers

Cultivar	Ferreira et al., (2016)	Silva et al., (2014)	Paz et al., (2015)	September-Malaterre et al., (2016)	Gardner et al., (2000)	Morais et al., (2015)	Fu et al., (2011)	Lu et al., (2014)
Imperial	126.95	-	329	-	358	-	-	-
Pérola	84. 90	-	-	-	-	197.87	-	56.84
IAC Fantástico	89.01	-	-	-	-	-	-	-
Smooth Cayenne	71.07	-	-	-	-	-	94.04	-
Gomo-de-mel	109.60	990.76	-	-	-	-	-	-
Thai Pulae	-	-	-	-	-	-	-	-
Nanglae	-	-	-	-	-	-	-	53.72
Vitória	74.09	-	-	33.0	-	-	-	-
Roja-española	-	-	-	-	-	-	-	-
Comte de Paris	-	-	-	-	-	-	-	48.01
CPM	-	-	-	-	-	-	-	31.48
Fresh Premium	-	-	-	-	-	-	-	56.21
Giant Kew	-	-	-	-	-	-	-	41.941
Kallara local	-	-	-	-	-	-	-	53.01
MacGregor	-	-	-	-	-	-	-	55.15
MD-2	-	-	-	-	-	-	-	77.55
New Puket	-	-	-	-	-	-	-	53.61
Pattavia	-	-	-	-	-	-	-	37.48
Phetchaburi #2	-	-	-	-	-	-	-	53.90
Puket	-	-	-	-	-	-	-	47.83
Queensland Cayenne	-	-	-	-	-	-	-	70.69
Ripley	-	-	-	-	-	-	-	54.20

-Not analyzed.

The presence of anthocyanins in pineapple pulp was detected. Silva et al., (2014) reported anthocyanins content (11.62 mg/100g) of fresh pulp but not detected the presence of yellow flavonoids. However, Paz et al., (2015) detected a low content (46 mg.EE/100g of fresh pulp). The low content can be explained since flavonoids are mainly present in plants as coloring pigments and the pineapple is fruit that does not have a strong color presence (Hossain & Rahman, 2011).

As presented by Fu et al. (2011), several phenolic compounds were identified by HPLC in pineapple such as: chlorogenic acid, luteolin- 7- O- glucoside, ferulic acid, protochatehuic acid and myricetin in the respective concentrations of 1.31 mg/100g; 3.72 mg/100g; 2.23 mg/100g; 4.11 mg/100g of fresh pulp and they suggested that these compounds are responsible for the biological and functional effects after the eating of this fruit.

Other not edible parts of pineapple are also rich source of phenolic compounds. Their contents in pineapple residues (seeds and peel) are reported to be greater than in flesh and could be a potential source for the extraction of beneficial bioactive compounds. In addition, diverse studies showed that phenolic content has a positive correlation with the antioxidant activity of the pineapple fruit (DPPH and FRAP assays) (Kongsuwan et al., 2009; Upadhyay et al., 2013).

In the work about phenolic compounds in pineapple peels, Li et al., (2014) identified four compounds viz gallic acid, catechin, epicatechin and ferulic acid by HPLC-MS and their concentrations were 31.76 mg/100g; 58.51 mg/100g; 50 mg/100g and 19.50 mg/100g of dried peel, respectively. Catechin and epicatechin are compounds when present in high amount in some fruits show a high antioxidant capacity, thus making pineapple residue to be a good source of extraction of these compounds.

2.3. Carotenoids

Carotenoids are classified chemically as polyisoprenoids. These are fat soluble compounds that are associated with the lipidic fractions and are divided into two main groups: (a) carotenes and (b) xanthophylls (Quirós & Costa, 2006). These are natural pigments which occur widely in nature and are synthesized by plants and many microorganisms, while animals have to obtain them from food. The content of carotenoids in foods may change depending on several factors such as, genetic variety, maturity, postharvest storage, processing and preparation (van den Berg et al., 2000).

Carotenoids have numerous beneficial effects on the human health such as antioxidants and enhancers of the immune response. Some good examples are β-carotene, α-carotene and cryptoxanthin which stand out for their provitamin A activity, which is converted into vitamin A or retinol after ingestion (Zeb & Mehmood, 2004). Due to their versatile health-promoting properties, the global market demand for carotenoids is estimated to increase from US$1.5 billion in 2014 to US$1.8 billion in 2019 (Strati & Oreopoulou, 2014).

Furthermore, some of them are involved in the cell communication and xanthophylls have shown to be effective as free radical scavengers showing antioxidant action, protecting cells and tissues from damage caused by free radicals, strengthening the immune system and inhibiting the development of certain types of cancers (Zeb & Mehmood, 2004; Krinsky, 1994).

The major carotenoids present in pineapple are violaxanthin (50%), luteoxanthin (13%), β- carotene (9%) and neoxanthin (8%), along with smaller amounts of hydroxyl- α- carotene, cryptoxanthin, lutein, auroxanthin and neochrome (Bauernfeind, 1981). Furthermore, pineapple by- products are also important sources of these compounds. Moreover, the presence of lutein and β- carotene in pineapple core has been reported (Freitas et al., 2015).

Some other studies also reported that pineapple contains carotenoids such as β-carotene, α-carotene and cryptoxanthin. Ferreira et al., (2016) determined the carotenoid content in the various pineapple cultivars and found that α-carotene showed the highest concentration, and it was higher in fruits of Gomo-de-Mel cultivar (498 mg/100g of pulp), followed by Imperial cultivar (365 mg/100g of pulp), while β-carotene showed higher concentration (226 mg/100g of pulp) in IAC Fantástico cultivar, but without significant differences from Gomo-de-Mel and Imperial cultivars while Zeaxanthin showed low concentration in all pineapple cultivars, even though in IAC Fantástico and Smooth Cayenne cultivars, their values were higher than the α-carotene concentration.

Setiawan et al., (2001) investigated the carotenoid content of some pineapple fruits, and quantified cryptoxanthin and β-carotene to be 89 and 230 mg/100 g, respectively. Viana et al., (2013) quantified only the total carotenoid content for Imperial cultivar and it was 266 μg/100g, while Smooth Cayenne, Pérola and Vitória cultivars showed lower concentrations, being 2.34; 0.69 and 0.32 μg/100g, respectively. However, Kongsuwan et al., (2009) found lower β-carotene levels for two Thai pineapple varieties, Phulae and Nanglae (3.35 and 1.41 μg/100 g, respectively), while another study with Thai fruits did not detect β-carotene in pineapple fruits (Charoensiri et al., 2009). These variations occur due to the same reasons as these are influenced by growing conditions, climate, cultivar, processing conditions. Table 4 assembles the data on carotenoid contents in pineapple, as reported by several researchers.

From the data reported until now, pineapple can be classified a good source of carotenoids, such as cryptoxanthin, β-carotene, lutein, α-carotene and zeaxanthin, when compared to the other fruits.

Table 4. Carotenoids (mg/100g of pulp) contents of pineapple according to various Researchers

Carotenoid	Ferreira et al., (2016)	Silva et al., (2014)	September-Malaterre et al., (2016)	Freitas et al., (2014)	Setiawan et al., (2001)	Hajare et al., (2006)
Lutein	20-250	-	-	288	-	-
Zeaxanthin	23-50	-	-	n.d.	-	-
Cryptoxanthin	45-275	-	-	n.d.	81	-
β-carotene	40-150	42.86	52.60	994-2537	230	658-778.9
α-carotene	15-480	-	-	89	-	-
Lycopene	-	n.d.	-	n.d.	399	-

-Not analyzed; n.d. = not detected.

2.4. Antioxidant Activity

The increase of consumption of food that can provide benefits like reducing the risk or to manage a specific disease is related to the compounds in food which are known as antioxidants. Natural antioxidants can be found in fruit and vegetables and have gained increasing interest among consumers and researchers. The epidemiological studies have indicated that frequent consumption of natural antioxidants is associated with a lower risk of cardiovascular disease and cancer (Renault et al., 1998). Antioxidants can be classified into primary and secondary, according to their mode of action. Primers act by interrupting the reaction chain by donating electrons or hydrogen to free radicals, converting them into thermodynamically stable products and/or reacting with free radicals, forming the lipid-antioxidant complex that can react with another free radical. However, the secondary antioxidants act by slowing the initiation stage of autoxidation, by different mechanisms that include metal complexation, oxygen sequestration, decomposition of hydroperoxides to form non-radical species, absorption of ultraviolet radiation or deactivation of singlet oxygen (Angelo & Jorge, 2007; Gonçalves, 2008).

Antioxidants participate as inhibitors of the reaction, donating hydrogen or receiving the free radicals of fatty acids, as well as in interfering in the action of singlet oxygen. When these act as receptors for free radicals 30 (AH), these react first with RO_2 and not with R • radicals, favoring a competition between antioxidants and the propagation of the chain reaction, in the presence of the fatty acid (RH), thus intervening in the reaction initiation phase, producing stable compounds that retard the oxidation process (Ribeiro & Seravalli, 2004).

The natural antioxidant capacity in fruit and vegetables are related to three major chemical groups; vitamins, phenolics and carotenoids. Ascorbic acid and phenolics are known as hydrophilic antioxidants, while carotenoids are known as lipophilic antioxidants (Kongsuwan et al., 2009; Halliwell, 1996). Due to a vast diversity of compounds possessing antioxidant capacity, the mechanisms of their action can be different requiring several methods for quantification of these compounds.

Many analytical studies involving DPPH and FRAP assays have shown that phenolic content has a positive correlation with the antioxidant activity of the pineapple fruit (Kongsuwan et al., 2009; Upadhyay et al., 2013). In particular, chlorogenic acid, luteolin- 7- O- glucoside, ferulic acid, protochatehuic acid and myricetin are individual phenolic compounds present in pineapple and are known to be responsible for the biological and functional effects after the eating of this fruit (Fu et al., 2011). Table 5 presents the data on the antioxidant capacity of various cultivars of pineapple.

2.5. Dietary Fiber (DF)

Dietary fiber is defined as non- starch polysaccharide, which resist hydrolysis by the endogenous enzymes of the digestive tract. They are carbohydrate polymers with ten or more monomeric units which are not hydrolyzed by endogenous enzymes in the small intestine of the humans (Kendall et al., 2010; Tosh & Yada, 2010). Dietary fibers such as cellulose, hemicellulose, lignin and pectin substances are a few examples (Valderrain-Rodriguez, 2017). The dietary fiber can be classified according to their solubility in water, as water- soluble and insoluble, and both have shown to have beneficial effects in the prevention of several diseases, such as cardiovascular diseases and diabetes (Rodríguez et al., 2006; Tosh & Yada, 2010).

The benefits of consumption of dietary fiber is related to their influence in various aspects of digestion, absorption and metabolism; among them: a) the reduction of intestinal transit time of food; b) increased intestinal absorption rate of glucose; c) decrease in blood cholesterol levels; and d) decrease in the content of calories ingested (Calixto, 1993; Botelho et al., 2002). These properties contribute to make the fiber an adequate intestinal regulator. Fibers are still important in dietary regimens for the prevention or treatment of diabetes, problems of hypercholesterolemia and obesity. On the contrary, lack of dietary fiber may be related to the development of colon cancer and other gastrointestinal disorders.

Another property of dietary fiber very much appreciated is its water retention capacity (WRC). From the physiological point of view, a higher WRC potentiates a larger volume of the food cake and, therefore, a greater sensation of satiety, greater volume and weight of the feces (Botelho et al, 2002). For these functions, dietary fiber has been indicated for the prevention of cardiovascular and gastrointestinal diseases (Silva et al., 1996).

The contribution of dietary fiber will be different depending on their type. Cellulose can act promoting the correct operation of human body correcting intestinal malfunction, such as constipation which is very common in hot climates (Calixto, 1993). Hemicellulose is known as a carbohydrate reserve and a potential source of sugars and other substances during fruit maturation (Soto, 1992). Lignin, due to its three-dimensional chemical structure and the presence of phenolic groups and hydrophobic properties, can act as an ion-exchange resin, binding to bile acids and hence contributing to the reduction of the formation of carcinogenic metabolites (Silva et al., 1996). Moreover, pectin substances function in the texture of the food and are recommended in human nutrition, since these have actions on digestion, usually in the intestinal functions (Botelho, et al., 2002).

Pineapple is a rich source of dietary fiber for the human diet. This fruit contains between 1 to 6.8% of total dietary fiber, depending on the pineapple variety and the part of the fruit being analyzed (USDA, 2014). According to Bartolomé & Rupérez (1995), pineapple is very rich in hemicellulose. Sgarbieri (1966) reported that the consistency of pineapple pulp depends on the content of certain carbohydrate constituents of high molecular weight, such as hemicellulose and pectins. Hemicelluloses are found in the cell wall, together with cellulose and lignin (Griswold, 1972). Table 6 presents data on the contents of dietary fiber of different cultivars of pineapple and in different parts of the fruit.

Table 5. Antioxidant capacity as determined by DPPH, FRAP, ABTS and TEAC assays in various cultivars of pineapple

Cultivar	FRAP	DPPH	ABTS	TEAC	References
Josapine	118.29	-	-	-	Chiet et al., (2014)
Morris	306.45	-	-	-	
Sarawak	73.92	-	-	-	
Comte de Paris		4.25		5.71	Lu et al., (2014)
CPM		3.68		4.10	
Fresh Premium		8.08		9.28	
Giant Kew		3.79		5.41	
Kallara local		5.60		7.04	
MacGregor		13.55		12.24	
MD-2		22.85		17.30	
Nanglae		6.84		6.85	
New Puket		9.32		7.80	
Pattavia		5.60		7.23	
Pearl		11.43		10.72	
Phetchaburi #2		8.44		7.63	
Puket		5.44		5.81	
Queensland Cayenne		14.76		14.24	
Ripley		9.99		13.11	

-Not analyzed.

Some dietary fiber can have antioxidant activity when combined with natural antioxidant compounds such as phenolic compounds. It has been reported that pineapple pulp, core and peel are good sources of dietary fiber combined with phenolic compounds that present antioxidant activity (Freitas et al., 2015). Larrauri et al. (1997) reported that pineapple fiber from shells exhibited a higher antioxidant activity (86%) than orange peel fiber (36.4%) due to presence of phenolic compounds (59%) such as myricetin, salicylic, tannic, trans-cinnamic and *p*-coumaric acids. Although the antioxidant dietary fiber content of pineapple pulp compared to other fruits is not high, its composition and presence of bound PC indicates that the consumption of this fruit may provide diverse and important health benefits (Valderrain-Rodriguez, 2017).

Pineapple residue is an excellent source of dietary fiber. Some publications reported that the content of dietary fiber in pineapple shells can be around 71% of dietary fiber, where the insoluble fraction constitute

99% of total dietary fiber ((Larrauri et al., 1997). Considering the large amount of industrially processed pineapple production, the residue generated should be considered as a good source for obtaining dietary fiber to be used as an ingredient for nutraceutical and functional food products (Valderrain-Rodriguez, 2017).

Table 6. Dietary fiber contents in pineapples of various cultivars

Cultivar	TDF (%)	SDF (%)	IDF (%)	Lignin (%)	Cellulose (%)	Hemicellulose (%)	Total Pectin (g/100g)	References
Red Spanish	1.09	-	-	-	-	-	-	Bartolomé et al., (1995)
Smooth Cayenne Pulp	1.63	-	-	1.33	2.70	4.18	0.32	Botelho et al., (2012)
Smooth Cayenne Core	-	-	-	0.66	0.95	1.12	0.27	Botelho et al., (2012)
Perola Pulp	-	0.02	0.78	-	-	-	-	Cordenunsi et al., (2010)
Perola Core	-	0.11	1.76	-	-	-	-	
Extra Sweet	1.40	-	-	-	-	-	-	USDA (2014)
Esmeralda	6.79	1.82	5.01	-	-	-	-	Velderrain - Rodríguez (2013)
Smooth Cayenne from Taiwan	-	-	-	-	12.58	6.77	2.13	Sun et al., (2016)
Smooth Cayenne from Hawaii	-	-	-	-	24.99	8.67	4.01	
Smooth Cayenne from Australia	-	-	-	-	14.44	13.00	2.41	
Comte de Paris	-	-	-	-	19.49	7.00	3.21	
Yuecui	-	-	-	-	22.64	15.03	4.15	
Zhenzhu	-	-	-	-	15.33	14.33	2.32	

TDF - Total Dietary Fiber; SDF - Soluble Dietary Fiber; ISF - Insoluble Dietary Fiber.
-Not analyzed.

2.6. Bromelain Enzyme

Bromelain (EC 3.4.22.32) belongs to a group of proteolytic enzymes and it is present in higher concentration in pineapple fruit. It is a bioactive macromolecule with good pharmacological activities (Bresolin et al., 2013). The name "bromelain" was first used to describe any plant member of the Bromeliaceae family, to which the pineapple belongs. It is used to describe proteolytic enzymes found in tissues, such as the stem, fruit and leaves, of the Bromeliaceae family. It has been investigated since 1884 (Devakate et al., 2009) and was first identified in 1891 by Marcano (Balls et al., 1941).

The enzyme has a lot of therapeutic efficacies like antiedemateous activity, antiinflammatory, antithrombotic and fibrinolytic activities. This enzyme can modulate the functions of adhesion molecules on blood and endothelial cells, and also regulate and activate various immune cells and their cytokine production as drugs for the oral systemic treatment of inflammatory, blood-coagulation-related and malignant diseases (Maurer, 2001). Other uses for bromelain commercially are in the food industry, in certain cosmetics and in dietary supplements (Uhlig, 1998; Walsh, 2002). Due to a large versatility of applications, commercial bromelain is very expensive costing up to 2400 US dollars/kg (Ketnawa et al., 2012). A summary on the pharmacological applications of bromelain is shown in the Table 7.

Bromelain is considered as a food supplement and is freely available to the general public in health food stores and pharmacies in the USA and Europe (Ley et al., 2011). The concentration of bromelain is high in pineapple stem, thus necessitating its extraction since, unlike the pineapple fruit which is normally used as food, the stem is a waste byproduct and thus it is inexpensive (Pavan et al., 2012).

The bromelain extract obtained from the stem and fruit bromelain was earlier (EC. 3.4.22.32) from fruit (EC. 3.4.22.33) are different and these are called as bromelin (Rowan, & Butle, 1994). The bromelain found in the pineapple stem has an isoelectric point (pI) of 9.5, and is the most abundant protease in pineapple tissue preparations, while the bromelain found in the pineapple fruit has a pI of 4.6, and is present in lesser amounts compared to the stem. The Table 7 shows the main differences between stem and fruit bromelain.

Table 7. Summary of pharmacological applications of bromelain

Applications	Disorders	References
Anti-Inflammatory agent	Asthma	Jaber et al., (2002); Secor et al., (2008)
	Chronic Rhinosinusitis	Bakhshaee et al., (2014)
	Colonic inflammation	Darshan; Doreswamy, (2004)
	Osteoarthritis	Klein (2006); Walker et al., (2002)
	Rheumatoid arthritis	Maurer, (2001)
	Soft tissue injuries	Baumann et al., (2007); Lemay et al., (2004)
Anti-tumor agent	Breast cancer	Baez et al., (2007)
	Leukemia	Maurer, (2001); Pavan et al., (2012)
	Lung carcinoma	QIMR, (2005)
	Ovarian cancer	Maurer, (2001)
	Melanoma	Kalra et al., (2007)
Improvement of Gastrointestinal Tract related discomforts	Postoperative gastrointestinal dysmotility (ileus)	Wen et al., (2006)
	Constipation	Wen et al., (2006)
	Exocrine pancreas insufficiency	Knill-Jones et al., (1970)
	Nausea, vomiting, diarrhea	Pavan et al., (2012)
Inhibition of thrombus formation	Angina pectoris	Taussig and Nieper, (1979)
	Transient ischemic attacks	Taussig and Nieper, (1979)
	Thrombophlebitis	Kelly, (1996)
	Thrombosis	Kelly, (1996)
Treatment of Dermatological disorders	Burns/Eschar	Rosenberg et al., (2004)
	Wrinkles	Reddy et al., (2013)
	Pityriasis lichenoides chronica	Massimiliano et al., (2007)
	Scleroderma	Gaby, (2006)

*Table adapted from Manzoor et al., (2016).

To obtain the enzyme from pineapple, various operations such as centrifugation, ultrafiltration and lyophilization are performed. The process yields a yellowish powder, the enzyme activity of which is determined with different substrates such as casein (FIP units), gelatine (gelatine digestion units) or chromogenic tripeptides (Maurer, 2001).

Ketnawa et al., (2012) reported for both Nang Lae and Phu Lae pineapple cultivars residues the content of bromelain as a major enzyme with the MW~28 kDa for the both cultivars. Umesh et al., (2008) reported in their study that the bromelain extracted from pineapple cores was found to be around 26 kDa. However, Maurer (2001) reported that the bromelain extracted from stems and fruits of pineapple were 23.8 kDa and 23 kDa, respectively. Therefore, pineapple fruit is very important source of bromelain since its high quantity found in the fruit has a great diversity of applications in food and cosmetic industries.

3. Health Promoting Factors of Bioactive Metabolites from Pineapple

In recent times, individuals prefer to control the risk of health ailments through consumption of functional foods. Phytotheraphy deals with an advanced variety of phytonutrients and enzymes as bioactive metabolites which conserve progression and metabolic rate of human physiological systems (Gupta and Prakash, 2014). Alarming research about enlightening health conditions, linking plant products with prospective human welfares, has heightened investigation on natural antioxidants (Kasote et al., 2015). Many worsening human diseases including cancer, inflammation, cardiovascular, cataracts, diabetes, Alzheimer's disease, macular degeneration and neurodegenerative diseases have been predictable as creatures of a probable concern of free radical injury to lipids, proteins and nucleic acids (Monroy et al., 2013). A preferential method to combat these diseases is to boost our body's immune system using natural antioxidants. Regular intake of natural foods such as fruits and vegetables has been related with a reduced frequency of worsening diseases (Pem and Jeewon, 2015). Preferentially fruits assist to boost immune system in copious ways. Pineapple is a globally obtainable fruit with numerous bioactive compounds like vitamin C, β-carotene, carotenoids, phenolic compounds and bromelain enzyme, which have been reported for high nutritive value and plentiful pharmacological benefits.

Oxidative stress has been evidenced to be the source of various diseases while numerous scientific investigations have been underway with issues such as free radicals mediated oxidative damages of human health condition. Fruits dominate in antioxidant defense principally due to the occurrence of phenolic compounds (Sytar et al., 2018). Flavonoids of polyphenol structure chiefly represent as coloring pigments in fruits and vegetables, and these act as antioxidant defensive role in neutralizing free radical damage in humans. Earlier studies evidenced that polyphenols derived from fruit species could protect lipid destruction from oxidation mechanism (Lin et al., 2016; Monsalve et al., 2017). Pineapple has an extraordinary nutritive content and a rich source of phenolic compounds. Although there are few reports on the antioxidant activities of pineapple in relation to the presence of phenolic compounds (Ferreira et al., 2016), the presence of vitamin C along with carotenoids is responsible for the synergistic defense responses against free radical mediated oxidative damage. It is quite popular that phenolic constituents, secondary metabolites from functional plant food substances protect themselves against various pathogens which authenticate the source of antimicrobial agent manufactures from plant species (Shin et al., 2018).

Bromelain, an enzyme derived from fruit or stem of pineapple which is commercially available at markets due to its enormous capabilities in aiding digestion. Bromelain obtained from fruit and stem of pineapple differed in composition of phytochemicals and biomolecules (Manzoor et al., 2016). It is a combination of various substances like thiol endopeptidases, phosphatase, glucosidase, peroxidase, cellulase, escharase, and is substantially digested in the human system without misplacing its proteolytic action and as well, being void of side effects (Pavan et al., 2012). Previous scientific reports evidenced that bromelain exhibited various pharmacological activities include fibrinolytic, antiedematous, antithrombotic, and anti-inflammatory activities. Bromelain has been well reported for potential health benefits like the management of bronchitis, sinusitis, surgical trauma, thrombophlebitis, wound healing, and enhanced absorption of drugs, particularly antibiotics. It is also used for curing osteoarthritis, diarrhea, and various cardiovascular ailments. Bromelain also possessed anticancerous activities and promotes apoptotic cell death (Pavan et al., 2012).

Conclusion

This chapter on the pineapple fruit focuses on its bioactive compounds present in the fruit along with their nutritional benefits. Although there are few publications about the bioactive compounds of pineapple, the main results are related to the vitamin C content of the fruit and the carotenoids. This review reveals the necessity of undertaking more research on the content of phenolic compounds such as flavonoids and other chemical substances. Overall, this review updates also the work done on pineapple, however, it indicates that there is a greater need to undertake extensive systematic research so as to explore its potential fully.

Acknowledgments

This study was financed in part by the Conselho Nacional de Desenvolvimento Científico e Tecnológico – Brasil (CNPq) vide projeto do INCT Tropical Fruits, the Fundação de Apoio à Pesquisa e a Inovação Tecnológica do Estado de Sergipe (Fapitec/SE) – Brasil and the Coordenação de Aperfeiçoamento de Pessoal de Nível Superior - Brasil (CAPES- Finance Code 001).

References

Alias, Nor H., and Zulkifly Abbas. 2017. "Preliminary Investigation on the Total Phenolic Content and Antioxidant Activity of Pineapple Wastes via Microwave Assisted Extraction at Fixed Microwave Power". *Chemical Engineering Transactions* 56:1675-1680.

Almeida, Maria M. B., Paulo H. M. Sousa, Maria L. Fonseca, Carlos E. C. Magalhães, Maria F. G Lopes, and Telma L. G. Lemos. 2009. "Evaluation of Macro and Micro-mineral Content in Tropical Fruits Cultivated in the Northeast of Brazil." *Ciência e Tecnologia de Alimentos* 29(3):581-586.

Almeida, Maria M. B., Paulo H. M. Sousa, Ângela M. C. Arriaga, Giovana M. Prado, Carlos E. C. Magalhães, Geraldo A. Maia, and Telma L. G. Lemos. 2011. "Bioactive Compounds and Antioxidant Activity of Fresh Exotic Fruits from Northeastern Brazil." *Food Research International* 44(7):2155-2159.

Báez, Roxana, Miriam T. P. Lopes, Carlos E. Salas, and Martha Hernandez. 2007. "*In vivo* Antitumoral Activity of Stem Pineapple (*Ananas comosus*) Bromelain." *Planta Medicinais* 73(13):1377-1383.

Bakhshaee, Mehdi, Farahzad Jabari, Mohammad M. Ghassemi, Shiva Hourzad, Russell Deutscher, and Kianoosh Nahid. 2014. "The Prevalence of Allergic Rhinitis in Patients with Chronic Rhino-sinusitis." *Iran Journal Otorhinolaryngology* 26(77): 245-249.

Bartholomew, Duane P., Robert E. Paul, and Kenneth G. Rorbach. 2003. *The pineapple: Botany, Production and Uses.* USA.: CABI Publishing.

Bartolomé, Ana P., Pilar Rupérez, and Carmen Fúster. 1995. "Pineapple Fruit: Morphological Characteristics, Chemical Composition and Sensory Analysis of Red Spanish and Smooth Cayenne Cultivars." *Food Chemistry* 53(1): 75–79.

Bauernfeind, J. Christopher. 1981. *Carotenoids as Colorants and Vitamin A Precursors: Technological and Nutritional Applications*. New York: Academic Press.

Baumann, Leslie S. 2007. "Botanical ingredients in cosmeceuticals." *Journal of Drugs in Dermatology* 6(11): 1084-1088.

Benassi, Marta de T., and Adriana Z. Mercadante. 2018. "Vitaminas." In Química e Bioquímica dos Alimentos, edited by Franco Maria Lajolo and Adriana Zerlotti Mercadante, 195-239. São Paulo: Editora Atheneu. ["Vitamins." In *Food Chemistry and Biochemistry*]

Benitez, Sheila, Luis Soro, Isabel Achaerandio, Francesc Sepulcre, and Montserrat Pujola. 2014. "Combined Effect of a Low Permeable Film and Edible Coatings or Calcium Dips on The Quality of Fresh-Cut Pineapple." *Journal of Food Process Engineering* 37: 91–99.

Botelho, Lidiane, Alzira da Conceição, and Vânia D. Carvalho. 2002. "Caracterização de Fibras Alimentares da Casca e Cilindro Central do Abacaxi 'Smooth Cayenne'." Ciência Agrotécnica, 26(2): 362-367. ["Characterization of Peeling Fibers and Central Smooth Cayenne Pineapple Cylinder." *Agrotechnic Science*]

Bresolin, Iara Rocha A. P., Igor T. L. Bresolin, Edgar Silveira; Elias B. Tambourgi, and Priscila G. Mazzola. 2013. "Isolation and Purification of Bromelain from Waste Peel of Pineapple for Therapeutic Application." *Brazilian Archives Biology Technology* 56(6):971–979

Calixto, Fulgêncio Saura; 1993. "Fibra Dietética de Manzana: Hacia Nuevos Tipos de Fibras de Alta Calidad." Alimentaria 4(1): 57-61. ["Apple Dietary Fiber: Towards New Types of High Quality Fibers." *Alimentaria*]

Charoensiri, R., R. Kongkachuicha, S. Suknicom, and P. Sungpuag. 2009. "Beta-carotene, Lycopene, and Alpha-Tocopherol Contents of Selected Thai Fruits." *Food Chemistry* 113:202–207.

Cordenunsi, Beatriz, Fulgêncio Saura- Calixto, Maria H. Diaz- Rubio, Angela Zuleta, Marco A. Tiné, Marcos S. Buckeridge, Giovanna B. da Silva, Cecilia Carpio, Eliana B. Giuntini, Elizabete W. de Menezes, and Franco Lajolo. 2010. "Carbohydrate Composition of Ripe Pineapple (cv. Perola) and the Glycemic Response in Humans." *Food Science and Technology* 30(1): 282–288.

Darshan S, and R. Doreswamy. 2004. "Patented Anti-Inflammatory Plant Drug Development from Traditional Medicine." *Phytotherapy Research* 18(5): 343-357.

Diplock, Anthony T. 1994. "Antioxidants and disease prevention." *Molecular Aspects of Medicine* 15(4): 293–376.

FAOSTAT, 2014. Accessed in http://www.fao.org/faostat/en/.

Ferreira, Ester Alice, Heloisa E. Siqueira, Eduardo V. Vilas Boas, Vanessa S. Hermes, and Alessandro O. Rios. 2016. "Bioactive Compounds and Antioxidant Activity of Pineapple Fruit of Different Cultivars." *Revista*

Brasileira de Fruticultura 38 (3): 146. [*Brazilian Journal of Fruit culture*]

Freitas, Ana, Margarida Moldão-Martins, Helena S. Costa, Tânia G. Albuquerque, Ana Valente, and Ana Sanches-Silva. 2015. "Effect of UV-C Radiation on Bioactive Compounds of Pineapple (*Ananas comosus* L. Merr.) by Products." *Journal Science Food Agriculture* 95: 44–52.

Fu, Li, Bo-Tao Xu, Xiang-Rong, Ren-You Gan, Yuan Zhang, En-Qin Xia, and Hua-Bin Li. 2011. "Antioxidant Capacities and Total Phenolic Contents of 62 Fruits." *Food Chemistry* 129(2): 345–350.

Gaby, Alan R. 2006. "Natural Remedies for Scleroderma." *Alternative Medical Review* 11(3): 188-195.

Gardner, Peter T., Tamsin A. C. White, Donald B. McPhail, and Garry G. Duthie. 2000. "The Relative Contributions of Vitamin C, Carotenoids and Phenolics to the Antioxidant Potential of Fruit Juices." *Food Chemistry* 68: 471-474.

Griswold, Ruth M.1972. *Estudo experimental dos alimentos.* São Paulo: Blucher. [*Experimental study of food*.]

Gupta, Charu, and Dhan Prakash. 2014. "Phytonutrients as Therapeutic Agents." *Journal of complementary and integrative medicine* 11(3): 151-169.

Hajare, Sachin N., Varsha S. Dhokane, R. Shashidhar, Sunil Saroj, Arun Sharma, Jayant R. Bandekar. 2006. "Radiation Processing of Minimally Processed Pineapple (*Ananas comosus* Merr.): Effect on Nutritional and Sensory Quality." *Journal of Food Science* 71: 501-505.

Hassimoto, Neuza M. A., and Adriana Z. Mercadante. 2018. "Pigmentos Naturais." In *Química e Bioquímica dos Alimentos*, edited by Franco Maria Lajolo and Adriana Zerlotti Mercadante, 159-239. São Paulo: Editora Atheneu. ["Natural Pigments." In *Food Chemistry and Biochemistry*]

Hossain, Md. Farid, Shaheen Akhtar, and Mustafa Anwar. 2015. "Nutritional Value and Medicinal Benefits of Pineapple." *International Journal of Nutrition and Food Sciences* 4(1): 84-88.

Jaber, Raja. 2002. "Respiratory and Allergic Diseases: From Upper Respiratory Tract Infections to Asthma." *Primary Care: Clinics in Office Practice* 29(2): 231-261.

Kasote, Deepak M., Surendra S. Katyare, Mahabaleshwar V. Hegde, and Hanhong Bae. 2015. "Significance of Antioxidant Potential of Plants and its Relevance to Therapeutic Applications." *International Journal of Biological Sciences* 11(8): 982-991.

Kendall, Cyrill W. C., Amin Esfahani, and David J. A. Jenkins. 2010. "The Link Between Dietary Fibre and Human Health." *Food Hydrocolloids* 24(1):42–48.

Ketnawa, Sunantha, Phanuphong Chaiwut, and Saroat Rawdkuen. 2012. "Pineapple Wastes: A Potential Source for Bromelain Extraction." *Food and Bioproducts Processing* 90: 385–391.

Kongsuwan, A., Phunsiri Suthiluk, Theerapong Theppakorn, Varit Srilaong, and Sutthiwal Setha. 2009. "Bioactive Compounds and Antioxidant Capacities of Phulae and Nanglae Pineapple." *Asian Journal of Food and Agro-Industry* 2(1): 44-50.

Krinsky, Norman I. 1994. "The Biological Properties of Carotenoids." *Pure and Applied Chemistry* 66(5): 1003-1010.

Kris- Etherton, Penny M., Kari D. Hecker, Andrea Bonanome, Stacie M. Coval, Amy E. Binkoski, Kirsten F. Hilpert, Amy E. Griel, and Terry D. Etherton. 2002. "Bioactive Compounds in Foods: Their Role in the Prevention of Cardiovascular Disease and Cancer." *American Journal of Medicine* 113(9):71–88.

Larrauri, José A., Pilar Rupérez, and Fulgencio Saura Calixto. 1997. "Pineapple Shell as a Source of Dietary Fiber with Associated Polyphenols." *Journal of Agricultural and Food Chemistry* 45(10): 4028–4031.

Ley, Chitmoy M., Amalia Tsiami, and Qing Ni, Nicola Robinson. 2011. "A review of the Use of Bromelain in Cardiovascular Diseases." *Journal of Chinese Integrative Medicine* 9 (7): 702-710.

Lin, Derong, Mengshi Xiao, Jingjing Zhao, Zhuohao Li, Baoshan Xing, Xindan Li, Maozhu Kong, Liangyu Li, Qing Zhang, Yaowen Liu, Hong Chen, Wen Qin, Hejun Wu, and Saiyan Chen. 2016. "An

Overview of Plant Phenolic Compounds and their Importance in Human Nutrition and Management of Type 2 Diabetes." *Molecules* 21: 1374.

Lu, Xin-Hua, De-Quan Sun, Qing-Song Wu, Sheg-Hui Liu, and Guang-Ming Sun. 2014. "Physico-Chemical Properties, Antioxidant Activity and Mineral Contents of Pineapple Genotypes Grown in China." *Molecules* 19: 8518-8532.

Manzoor, Zoya, Ali Nawaz, Hamid Mukhtar, and Ikram Haq. 2016. "Bromelain: Methods of Extraction, Purification and Therapeutic Applications." *Brazilian Archives of Biology and Technology* 59.

Massimiliano, Risulo, Pietro Rubegni, Paolo Sbano, Sara Poggiali, and Michele Fimiani. 2007. "Role of Bromelain in the Treatment of Patients with Pityriasis Lichenoides Chronica." *Journal Dermatology Treatment* 18(4): 219-222.

Maurer, H. Rainer. 2001. "Bromelain: Biochemistry, Pharmacology and Medical Use." *Cellular and Molecular Life Sciences* 9: 1234–1245.

Monroy, Adriana, Gordon J. Lithgow, and Silvestre Alavez. 2013. "Curcumin and Neurodegenerative Diseases." *Biofactors* 39(1): 122-132.

Monsalve, Bernardita, Anibal Concha-Meyer, Iván Palomo, and Eduardo Fuentes. 2017. "Mechanisms of Endothelial Protection by Natural Bioactive Compounds from Fruit and Vegetables." *Annals of the Brazilian Academy of Sciences* 89 (supplement).

Morais, Damila R., Eliza M. Rotta, Sheisa C. Sargi, Eduardo M. Schmidt, Elton G. Bonafe, Marcos N. Eberlin, Alexandra C. H. F. Sawaya, and Jesuí V. Visentainer. 2015. "Antioxidant Activity, Phenolics and UPLC–ESI(–)–MS of Extracts from Different Tropical Fruits Parts and Processed Peels." *Food Research International* 77: 392–399.

Naczk, Marian, and Fereidoon Shahidi. 2004. "Extractions and Analysis of Phenolics in Food." *Journal of Chromatography A* 1054: 95- 111.

Pavan, Rajendra, Sapna Jain, Shraddha, and Ajay Kumar. 2012. "Properties and Therapeutic Application of Bromelain: A Review." *Biotechnology Research International* 1: 1-6.

Paz, Mário, Patricia Gúllon, Maria F. Barroso, Ana P. Carvalho, Valentina F. Domingues, Ana M. Gomes, Helena Becker, Elisane Longhinotti and Cristina Delerue-Matos. 2015. "Brazilian Fruit Pulps as Functional Foods and Additives: Evaluation of Bioactive Compounds." *Food Chemistry* 172: 462–468.

Pem, Dhandevi, and Rajesh Jeewon. 2015. "Fruit and Vegetable Intake: Benefits and Progress of Nutrition Education Interventions - Narrative Review Article." *Iranian Journal of Public Health* 44 (10): 1309-1321.

Quirós, Ana R.-B., and Helena S. Costa. 2006. "Analysis of Carotenoids in Vegetable and Plasma Samples: A review." *Journal of Food Composition and Analysis* 19: 97–111.

Rocha, Marina Sousa, 2011. "Compostos Bioativos e Atividade Antioxidante (*in vitro*) de Frutos do Cerrado Piauiense." Dissertação de mestrado, Universidade Federal do Piauí. ["Bioactive Compounds and Antioxidant Activity (*in vitro*) of Fruits of Cerrado Piauiense."]

Rodríguez, Rocío, Ana Jimenez, Juan Fernández- Bolaños, Rafael Guillén, and Antonia Heredia, 2006. "Dietary Fibre from Vegetable Products as Source of Functional Ingredients." *Trends in Food Science & Technology* 17(1): 3–15.

Rotili, Maria C. C., Sidiane Coutro, Viviane M. Celant, Jessica A. Vorpagel, Fabine K. Barp, Ariane B. Salibe, and Gilberto C. Braga. 2013. *Composição, Atividade Antioxidante e Qualidade do Maracujá-Amarelo Durante Armazenamento.* Semina: Ciência Agrária. 34 (1): 227-240. [*Composition, Antioxidant Activity and Quality of Yellow Passion Fruit during Storage*.]

Rowan, Andrew D., and David J. Buttle. 1994. "Pineapple Cysteine Endopeptidases." *Methods in Enzymology* 244: 555–568.

Septembre-Malaterre, Axelle, Giovédie Stanislas, Elisabeth Douraguia, and Marie-Paule Gonthier. 2016. "Evaluation of Nutritional and Antioxidant Properties of the Tropical Fruits Banana, Litchi, Mango, Papaya, Passion Fruit and Pineapple Cultivated in Réunion French Island." *Food Chemistry* 212: 225–233.

Setiawan, Budi, Ahmad Sulaeman, David W. Giraud, and Judy A. Driskell. 2001. "Carotenoid Content of Selected Indonesian Fruits." *Journal of Food Composition and Analysis* 14(2):169-176.

Sgarbieri, Valdemiro Carlos. 1966. "Estudo da composição química do abacaxi." *Boletim do Centro Tropical de Pesquisa e Tecnologia de Alimentos* 7(1): 37-50. ["Study of the chemical composition of pineapple." *Bulletin of the Tropical Center for Research and Technology of Food*]

Shin, Jonghoon, Vasantha-Srinivasan Prabhakaran, and Kwang-sun Kim. 2018. "The Multi-Faceted Potential of Plant-Derived Metabolites as Antimicrobial Agents Against Multidrug-Resistant Pathogens." *Microbial Pathogenesis* 116: 209-214.

Sies, Helmut, and Wilhelm Stahl. 1995. "Vitamins E and C, β-carotene, and other Carotenoids as Antioxidants." *American Journal of Clinical Nutrition* 62(6): 1315-1321.

Silva, M. E. B. da, C. M. S. da Silva, and L. T. Seara. 1996. "Influência da dieta na incidência e desenvolvimento do câncer de cólon e reto em Maceió." In: *Congresso Brasileiro de Ciência e Tecnologia de Alimentos.* ["Influence of diet on the incidence and development of colon and rectum cancer in Maceió." In: *Brazilian Congress of Food Science and Technology*.]

Silva, Mara R., Diracy B. C. L. Lacerda, Grazielle G. Santos, and Denise M. de O. Martina. 2008. "Caracterização Química dos Frutos nativos do cerrado." *Ciência Rural* 38(6): 1790-1793. ["Chemical Characterization of native fruits of the cerrado." *Rural Science*]

Silva. Larissa M. R., Evania A. T. Figueiredo, Nagila M. P. S. Ricardo, Icaro G. P. Vieira, Raimundo W. Figueiredo, Isabella M. Brasil, and Carmen L. Gomes. 2014. "Quantification of Bioactive Compounds in Pulps and by-Products of Tropical Fruits from Brazil." *Food Chemistry* 143: 398–404.

Soto, Ballestero M. 1992. *Bananos: cultivo e comercializacion.* Costa Rica: LIL. [*Bananas: cultivation and marketing*]

Sun, Guang-Ming, Xiu-Mei Zhang, Alain Soler, and Paul-Alex Marie-Alphonsine. 2016. "Nutritional Composition of Pineapple (*Ananas comosus* (L.) Merr.)." In *Nutritional Composition of Fruit Cultivars*, edited by Monique S. J. Simmonds and Victor R. Preedy, 609-637. London: Academic Press.

Sytar, Oksana, Irene Hemmerich, Marek Zivcak, Cornelia Rauh, and Marian Brestic. 2018. "Comparative Analysis of Bioactive Phenolic Compounds Composition from 26 Medicinal Plants." *Saudi Journal of Biological Sciences* 25(4): 631-641.

Taussig, Steven J., Joseph Szekerczes, and Stanley Batkin. 1985. "Inhibition of Tumour Growth *in vitro* by Bromelain, an Extract of the Pineapple Plant (*Ananas comosus*)." *Planta Medica* 51(6): 538-539.

Tosh, Susan M., and Sylvia Yada. 2010. "Dietary Fibres in Pulse Seeds and Fractions: Characterization, Functional Attributes, and Applications." *Food Research International* 43(2):450–460.

Uhlig, Helmut. 1998. *Industrial Enzymes and their Applications*, vol. 2. New York: John Wiley & Sons.

UNCTAD [United Nations Conference on Trade and Development] (2012) *INFOCOMM - Commodity Profile Pineapple. Pineapple: Ananas comosus (L.) Merr. of the Bromeliaceae family*, [15 Jan 2015].

Upadhyay, Atul, Jeewan P. Lama, and Shinkichi Tawata. 2013. "Utilization of Pineapple Waste: a Review." *Journal of Food Science and Technology* 6: 10–18.

USDA (United States Department of Agriculture). 2014. *National Nutrient Database for Standard Reference Release* 27, [Online], Available: http://www.ars.usda.gov/ba/bhnrc/ndl [20 Dec 2014].

Valderrain- Rodríguez, Gustavo R., Begoña de Ancos, Concepcíon Sánchez- Moreno, and Gustavo A. González- Aguilar. 2017. "Functional Properties of Pineapple." In: *Handbook of Pineapple Technology: Production, Postharvest Science, Processing and Nutrition*, edited by María Gloria Lobo and Robert E. Paull, 240-257. John Wiley & Sons.

Velderrain- Rodríguez, Gustavo Ruben. 2013. *Efecto de la fibra dietaria en la capacidad antioxidante de compuestos fenólicos de frutos tropicales durante un modelo de digestión in vitro.* Maestria en Ciencias, Centro de Investigación en Alimentación y Desarrollo.

[*Effect of dietary fiber on the antioxidant capacity of phenolic compounds of tropical fruits during an in vitro digestion model.* Master of Science, Center for Research in Food and Development]

Viana, Eliseth de Souza, Ronielli C. Reis, Jaciene L. de Jesus, Davi T. Junghans, and Fernanda V. D. Souza. 2013. "Caracterização Físico-Química de Novos Híbridos de Abacaxi Resistentes à Fusariose." *Ciência Rural* 43(7): 1155-1161. ["Physical-Chemical Characterization of New Fusarium Resistant Pineapple Hybrids." *Rural Science*]

Walker, Ann F., Rafe Bundy, Stephen M. Hicks, and Richard W. Middleton. 2002. "Bromelain Reduces Mild Acute Knee Pain and Improves Well-Being in a Dose-Dependent Fashion in an Open Study of Otherwise Healthy Adults." *Phytomedicine* 9(8): 681-686.

Walsh, Gary. 2002. *Protein: Biochemistry and Biotechnology*. New York: John Wiley & Sons.

Yeoh Wei K., and Asgar Ali. 2017. "Ultrasound Treatment on Phenolic Metabolism and Antioxidant Capacity of Fresh-Cut Pineapple During Cold Storage." *Food Chemistry* 216: 247–253.

Zeb, Alam, and Sultan Mehmood. 2004. "Carotenoids Contents from Various Sources and Their Potential Health Applications." *Pakistan Journal of Nutrition* 3(3): 199-204.

In: The Pineapple
Editor: Lydia Hampton
ISBN: 978-1-53614-594-6

Chapter 4

PINEAPPLE PEEL FLOUR AS A SOURCE OF FUNCTIONAL INGREDIENTS IN COOKED MEAT PRODUCTS

M. Lourdes Pérez-Chabela[1,*] and Alfonso Totosaus[2]
[1]Biotechnology Department, Universidad Autónoma Metropolitana, Ciudad de México, Mexico
[2]Food Science Lab and Pilot Plant, Tecnológico de Estudios Superiores de Ecatepec, Ecatepec, Mexico

ABSTRACT

Pineapple is one of the major tropical fruits produced in tropical and subtropical regions and consumed worldwide. As a consequence, during processing the peel represents a source of organic waste that can be employed as an important source of added value functional food ingredients. Pineapple peel flour contains a high amount of soluble and insoluble fiber, as well as polyphenols with antioxidant activity. The incorporation of pineapple peel flour into cooked meat products enhanced moisture and texture, and improved probiotic survival during storage.

[*] Corresponding Author Email: lpch@xanum.uam.mx.

Similarly, the antioxidant capacity of pineapple peel flour enhanced lipid rancidity in cooked meat products. Pineapple co-products are low-cost, have low caloric content and can be employed as a fiber source in many food products.

1. INTRODUCTION

1.1. Pineapple Production in the World

Pineapple is a native fruit of South America, growing in tropical zones, and is the second harvest of economic importance after bananas. Around 24.8 million tons are produced annually, where the top ten pineapple producers are Costa Rica, Brazil, Philippines, China, India, Thailand, Nigeria, Indonesia, Mexico and Colombia. Global sales from pineapple exportation were over US$2 billion in 2017. Nonetheless, since it is consumed fresh without the peel, the generation of agro-industrial co-products is an important environmental issue.

The edible portion is around 29-40% of total pineapple weight (Choonut et al., 2014), and is commonly eaten as canned fruit or juice, while the remainder is usually discarded after the manufacturing process (Damasceno et al., 2016). Pineapple peels, stems, crowns and cores are considered waste by the fruit pulp industry. These co-products contain a great amount of pectin and insoluble fiber, besides a high content of crude fiber and other compounds. The discarded fruit portion is a source of potential nutrients for diet supplementation, at low cost. The common practice was to use peel as cattle feed, but the disposal of this organic waste still constitutes a pollution problem.

1.2. Nutritive Value of Pineapple Peel

Table 1 shows the chemical composition of pineapple peel reported by various authors, with large differences due to the variety, climate changes, and other factors (Emaga et al., 2007). Pineapple peel is also a good source of vitamins and minerals. The main vitamin seems to be vitamin C, in a range from 35 to 89 mg/g of sample (Ramirez and Pacheco de Delahaye, 2011; Ferreira et al., 2016). Morais et al., (2017) reported that fruits like pineapple exhibited essential fatty acids as omega-6 and omega-3 with the largest contents observed in the peels.

Similarly, minerals reported in pineapple are calcium, zinc and iron, as well as micro-nutrient minerals such as manganese and sodium. Table 2 shows the mineral content reported by some authors. Nonetheless, peels can contain some antinutrients such as oxalates 129.06 mg/100 g, hydrogen cyanides 71.50 mg/100 g, alkaloids 16.19 mg/100 g and phytates 1.99 mg/100 g (Dibanda et al., 2016). In most cases these are water soluble and can be washed away by soaking in water.

Table 1. Chemical composition of pineapple peel

Moisture %	Protein %	Lipids %	Ashes %	Carbohydrates %	Reference
11.57	0.32	0.17	2.81	22.59	Diaz-Vela et al., 2013
9.3	4.00	1.30	4.5	14.40	Martínez et al., 2012
8.8	7.30	1.30	5.10	-	Morais et al., 2017
-	5.11	5.31	4.39	55.52	Dibanda et al., 2016
-	10.00	-	0.04	9.75	Hemalatha and Anbuselvi, 2013
-	6.27	0.10	1.14	77.19	Ramírez and Pacheco de Delahaye, 2011
-	0.50	0.20	-	13.10	Fischer, 2012

Table 2. Mineral content in pineapple peel (mg/100 g dry peel)

Mineral					Reference
Ca	Zn	Fe	Mn	Na	
8.30	6.46	25.52	5.32	-	Dibanda et al., 2016
64.60	0.80	1.60	8.20	9.80	Morais et al., 2017
6.95	-	4.20	-	-	Ramírez and Pacheco de Delahaye, 2011
2.80	8.97	2.80	2.45	-	Teixeira-Souza et al., 2016
5.40	2.30	15.30	2.27	-	Li et al., 2014

Pineapple peel is a potential source for bromelain extraction. This enzyme could be recovered and purified to generate a product with a higher added value. Bromelain is a generic name given to proteolytic enzymes (Bressolin et al., 2013). The different methods of extracting bromelain include liquid-liquid extraction by aqueous two-phase system, membrane filtration, precipitation methods, chromatographic processes (Novaes et al., 2016).

2. Functional Ingredients in Pineapple Peel

Pineapple peel is rich in functional ingredients including dietetic fiber with prebiotic potential and polyphenols with antioxidant activity.

2.1. Fiber

Fruit by-products are generated in large amounts, but commercial production of fruit fibers is limited since fresh fruit tissue after squeezing is not stable against enzymatic degradation, is sensitive to microbiological spoilage, and needs a drying process to reduce moisture content (Fisher, 2012). However, when the fresh fruit tissue or peel is stabilized, the fiber can be extracted and/or refined. Fiber is defined as "the remnants of plant cells resistant to the alimentary enzymes of man" (Trowell, 1977). In 2001, the American Association of Cereal Chemists (AACC, 2001) expanded the definition of fiber to edible parts of plants or analogous carbohydrates that are resistant to digestion and absorption in the human small intestine with complete or partial fragmentation in the large intestine, including polysaccharides, oligosaccharides, lignin and associated plant substances. Beneficial effects are attributed to fiber, such as reduced blood total and/or bad cholesterol levels, attenuation of postprandial glycemia, reduced blood pressure, increased fecal bulk, increased colonic fermentation, among others (Howlett et al., 2012).

Dietary fiber can be classified by its solubility, as insoluble dietary fiber, less fermented fiber, soluble dietary fiber, and well fermented fiber. Insoluble fiber consists mainly of cellulose, with a lesser amount of hemicellulose and lignin; soluble fiber consists of pentosans, pectins, gums and mucilage.

Table 3. Content of total fiber and soluble/insoluble fractions for pineapple peel

Dietary fiber			**Reference**
Soluble	Insoluble	Total	
21.66	40.88	62.54	Díaz-Vela et al., 2013
0.6	75.2	75.8	Martínez et al., 2012
5.90	36.3	42.2	Huang et al., 2011
33.67	46.20	81.8	Sánchez-Pardo et al., 2014
0.51	70.10	70.61	Larrauri et al., 1997
-	-	9.23	Teixeira-Souza et al., 2016

Soluble fiber not only performs certain important physiological functions, but also builds up important microflora by acting as a substrate food for beneficial microorganisms, acting therefore as a prebiotic to improve host health (Chawla and Patil, 2010). In contrast, insoluble fibers with higher viscosity provide bulk, increasing the rate of transit through the small bowel (Burton-Freeman et al., 2017). Huang et al., (2014) reported high levels of water-insoluble fiber-rich fraction in pineapple peel, with cellulose, hemicellulose (xylan and xyloglucan), and pectic substances as major polysaccharides. The amount of fiber varies depending on the cultivar, time of year, state of maturation, etc. Table 3 shows the fiber values reported by different authors.

2.2. Prebiotic Potential

A prebiotic is defined as "a selectively fermented ingredient that allows specific changes, both in the composition and/or activity in the gastrointestinal microflora that confers benefits upon host well-being and health" (Roberfroid, 2007). Prebiotics increase the number of beneficial bacteria (probiotics) in the gut. According to Gibson and Roberfroid (1995), criteria for classification of prebiotics are:

a) Resistance to digestive processes in the upper part of the gastrointestinal tract
b) Fermentation by intestinal microbiota
c) Selective stimulation of growth and/or activity of a limited number of the health-promoting bacteria in that microbiota.

Prebiotic activity of pineapple peel has been studied. Pyar et al., (2014) employed pineapple waste as carbon source in a fermentation media for three probiotic strains, with the best growth at 37 °C and pH 6. In a more complete study, Parra-Matadamas et al., (2015) employed pineapple peel flour as carbon source as well as two lactic acid bacteria strains (*P. pentosaceus* UAM21 and *A. viridans* UAM22), reporting a higher diauxic growth, where bacteria used the second carbohydrates, soluble fiber present in pineapple peel flours, after depletion of the first substrate, soluble simple carbohydrate. Pineapple peel powder can also be used in the dairy industry as a fiber source. Sah et al., (2016) used pineapple peel powder in yogurt, as compared to prebiotic inulin, reporting that syneresis and texture in both yogurts were similar during storage.

2.3. Polyphenols with Antioxidant Activity

Pineapple by-products from fruit processing provide valuable sources for the recovery of functional compounds like polyphenolic compounds, which are considered major contributors to the antioxidant activity of vegetables and fruits. Polyphenolic compounds are secondary plant metabolites that exist ubiquitously in the plant kingdom where they have a wide range of different structures and physiological properties, including anti-allergenic, anti-atherogenic, anti-inflammatory, anti-microbial, antioxidant, anti-thrombotic, cardio protective, and vasodilator effects. Interest in polyphenolic antioxidants has increased remarkably in the last decade because of their capacity to scavenge free radicals associated with various diseases.

Different polyphenols have been found in pineapple peel. Ribeiro da Silva et al., (2014) reported total phenolic content of 2787.09 mg/100 g of gallic acid equivalent (GAE), in addition to anthocyanins (10.10 mg/100g) and β-carotene (156 μg/100 g). Also, Li et al., (2014) reported catechin (50.51 mg/100 g), epicatechin (50.00 mg/100 g), gallic acid (31.76 mg/g) and ferulic acid (19.50 mg/g). Steingass et al., (2015) reported more than 60 phenolic compounds from different pineapple tissues. Ethanol-water (55:45) extracts from pineapple peel presented higher antioxidant activity in the range of 0.8 ± 0.05 to 1.3 ± 0.09 mg/mL, with a total phenolic content of 0.9 mg/g of GAE (Saraswaty et al., 2017). Similarly, fermentation of pineapple waste with *Kluyveromyces marxianus* NRRL Y-8281 increased the content of phenolic compounds (up to 120 mg/100 g of gallic acid, for example) (Rashad et al., 2015).

3. Functional Foods

Two thousand years ago, Hippocrates said "Let food be thy medicine and medicine be thy food." This is the principle of the functional foods definition. The term "functional food" was first employed in Japan in 1984, defined as food products fortified with special constituents that possess advantageous physiological effects (Kuvomara, 1998). Functional foods must exert a health or physiological effect, be part of ordinary foods (not as a supplement like pills or capsules) and be consumed as part of an ordinary diet (Howlett, 2008).

3.1. Functional Meat Products

The use of functional bioactive compounds in meat production can enhance human health. The choice of substance type and its quantity depends on the product meant to be obtained, more specifically, whether the product will be processed and if so, what kind of processing will be applied; cooked meat products can be fortified with fiber that improve water retention and texture, whereas in dry fermented meat products like salami, the use of soluble fiber can enhance texture and serve as a prebiotic. Equally important in functional meat product design is the consideration of dietary deficiencies of specific substances noted in various human populations. The process of creating new meat products with functional properties is complex and depends not only on the impact of applied functional ingredients on nutritional value, but also on the final quality of the meat (Pogorzelska-Nowicka et al., 2018). Nonetheless, meat and meat products are generally recognized as good sources of high biological value proteins, fat soluble vitamins, minerals, trace elements and bioactive compounds (Mehta et al., 2015).

Functional foods are an excellent opportunity for the meat industry to improve the quality and image of meat, not only to prevent the loss of market share attributable to a negative perception of meat and meat products, but by the development of products with health-beneficial properties (Olmedilla-Alonso et al., 2013).

The potential of using by-products as a source of functional ingredients in meat products has been studied. Freeze dried pineapple peel and pomace was employed in 1.5%, according to the total dietary fiber of pineapple by-product (64%), corresponding to the addition of 1.04% to the formulation of reduced fat beef burgers. Added pineapple improved yield with no detrimental effect on color, and increased hardness and cohesiveness, but with sensorial difference (Selani et al., 2016b). In this way, both the fatty acids profile and fat content of beef burgers can be modified. With the incorporation of canola oil, the polyunsaturated/saturated fatty acids ratio increased, cholesterol decreased, and oxidative stability was enhanced (Selani et al., 2016a). Also, commercial dietary fibers from a pineapple processing facility have been employed as fat replacers in sausages, presenting a good stability of added water since cooking loss and total expressible fluid values were close to control, although the sausages had a softer and less cohesive texture (St. Clair Henning et al., 2016).

An interesting approach was the study of pineapple peel flour as a potential functional ingredient in cooked meat products inoculated with thermotolerant lactic acid bacteria. Pineapple peel flour reduced expressible moisture, forming a harder and less cohesive structure, compared to control.

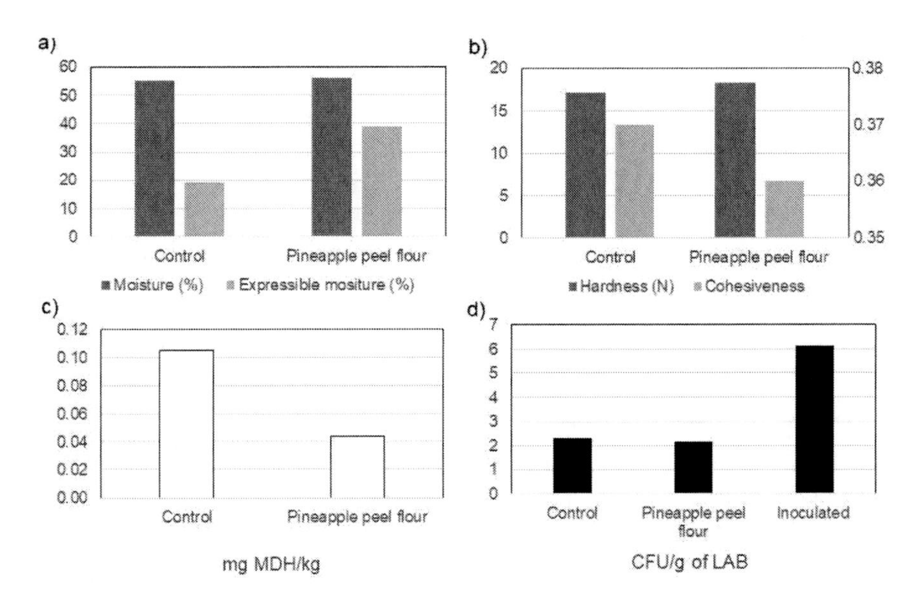

Figure 1. Effect of pineapple peel flour on cooked sausages: (a) moisture and expressible moisture, (b) texture, (c) oxidative rancidity, and (d) lactic acid bacteria growth.

In addition to decreasing oxidative rancidity during storage, microstructural analysis showed that the inoculated thermotolerant lactic acid bacteria produced exopolysaccharides, explaining textural differences (Díaz-Vela et al., 2015). The prebiotic activity of pineapple peel flours was tested *in vitro* (Díaz-Vela et al., 2013). The acceptance of functional meat products was also important. Functional meat products, such as cooked sausages with added dietary fiber, were acceptable mainly by middle-aged women, despite an apparent neophobia towards cooked sausages with fiber. In the sensory evaluation, sausages with added fiber were evaluated as acceptable, particularly the taste in pineapple fiber samples. The sensory analysis of new products is essential for subsequent acceptance by consumers, moreover in the functional food market (Díaz-Vela et al., 2017). The use of fruit peel from pineapple as a source of bioactive compounds (fiber, antioxidants and prebiotics) enhanced the development of thermotolerant lactic acid bacteria in cooked sausages during storage, being a viable alternative in the search for symbiotic meat products.

The incorporation of pineapple peel flour increases moisture in cooked meat sausages, but the water released in samples with pineapple peel flour could not be physically retained inside the product. Since the fiber contained is mostly insoluble, water cannot be chemically bonded and was released when force was applied (Figure 1a). However, the water retained by the fiber in peel flour made sausages tougher but less cohesive (Figure 1b); the compound in pineapple peel flour enforces the structure of the meat batter, but with a less consistent internal structure, resulting in a more easily breakable texture. At this point, the fiber present in peel is responsible for the improved texture of cooked sausages. The functional ingredients like polyphenols also had an important role in stabilizing the fat in cooked sausages. When pineapple peel flour was incorporated into cooked sausages, malonaldehyde (MDH) content was lower than in control sausages, improving oxidative rancidity (Figure 1c). Finally, fermentable carbohydrates in peel flour had prebiotic activity acting in synergy when thermotolerant probiotic lactic acid bacteria were inoculated into cooked sausages (Figure 1d). Thermotolerant lactic acid bacteria were inoculated during the elaboration of the sausages, surviving thermal processing to enhance nutritional characteristics of cooked sausages. This is probably the innovative feature to develop functional meat products. The soluble fiber in pineapple peel flour can be employed by probiotic thermotolerant lactic acid bacteria.

These kinds of foods are called symbiotic. The advantage of thermotolerant lactic acid bacteria is their inherent ability to survive sausage temperatures during thermal processing, and since they are inoculated before casing, serve as a bioprotective culture due to their antimicrobial properties, such as low pH and bacteriocin production. Under storage conditions, where sausages are vacuum packed and refrigerated, lactic acid bacteria become dominant flora. The fermentable sugars and soluble fiber present in pineapple peel flour have prebiotic potential, enhancing the growth of lactic acid bacteria. Additionally, insoluble fibers reinforce the meat batter structure enhancing textural properties (Figure 2).

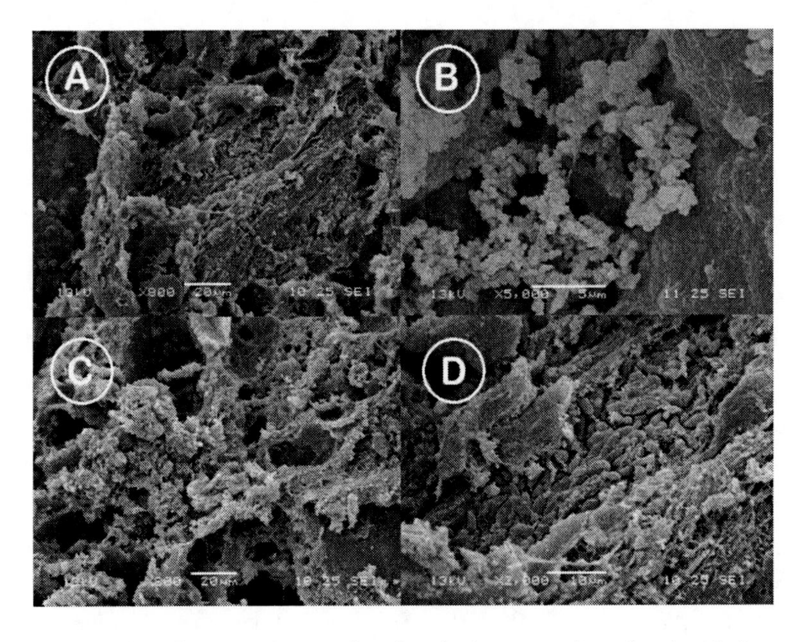

Figure 2. Scanning electron micrographs of cooked sausages inoculated with *P. pentosaceus* UAM22 and pineapple peel flour. A) day 1, where some fiber can be observed, with scarce presence of bacterial cells. B, C and D) at day 9 of storage (vacuum package, 4°C) lactic acid bacteria were widely dispersed within protein matrix, as well as pineapple peel fibers.

CONCLUSION

In accordance with this brief review, the main functional ingredients of pineapple co-products are antioxidants and fiber. Antioxidants could improve fatty acid stability during storage, and when soluble and insoluble fiber is incorporated into cooked meat products their hygroscopic properties enhance yield and texture. In most cases, fiber is employed as a fat replacer with added water, but fiber obtained from pineapple peel can also be employed as a prebiotic if thermotolerant lactic acid bacteria are inoculated. Pineapple peels are interesting as a nutritional ingredient for use in a range of commercial food-products. These co-products are inexpensive, non-caloric and may replace fat and flour in meat products as well as being a good functional ingredient in other food products.

REFERENCES

AACC. 2001. American Association of Cereal Chemist. Definition of dietary fiber. Report of the dietary fiber definition committee to the board of directors of the American Association of Cereal Chemist. *Cereal Foods World*, 46(3):112-126.

Bressolin, I. R. A. P., Lazarotto, B. I. T., Silveira, E., Tambourgi, E. B., and Mazzola, P. G. 2013. Isolation and purification of bromelain from waste peel of pineapple for therapeutic application. *Brazilian Archives of Biology and Technology* 56: 971-979.

Burton-Freeman, B. B., Liyanage, D., Rahman, S., and Edirisinghe, I. 2017. Radios of soluble and insoluble dietary fibers on satiety and energy intake in overweight pre- and post-menopausal women. *Nutrition and Healthy Aging* 4: 157-168.

Chawla, R., and Patil, G. R. 2010. Soluble dietary fiber. *Comprehensive reviews in Food Science and Food Safety,* 9: 178-196.

Choonut, A., Saejong, M., and Sangkharak, K. 2014. The production of ethanol and hydrogen from pineapple peel by *Saccharomyces cerevisiae* and *Enterobacter aerogenes*. *Energy Procedia*, 52: 242-249.

Damasceno, K. A., Alvarenga, G. C. A., Dos Santos, P. G., Lacerda, C. L., Bastianello, C. P. C., Leal de Ameida, P., and Arantes-Pereira, L. 2016. Development of cereal bars containing pineapple peel flour (*Ananas comosus* L. Merril) *Journal of Food Quality* 39: 417-424.

Díaz-Vela, J., Totosaus, A., Cruz-Guerrero, A. E., and Pérez-Chabela, M. L. 2013. *In vitro* evaluation of the fermentation of added-value agroindustrial by-products: Cactus pear (*Opuntia ficus indica* L.) peel and pineapple (*Ananas comosus*) peel as functional ingredients. *International Journal of Food Science and Technology* 48: 1460-1467.

Díaz-Vela, J., Totosaus, A., Escalona-Buendía, H. B., and Pérez-Chabela, M. L. 2017. Influence of the fiber from agroindustrial co-products as functional food ingredient on the acceptance, neophobia, and sensory characteristics of cooked sausages. *Journal of Food Science and Technology* 54: 379-385.

Díaz-Vela, J., Totosaus, A., and Pérez-Chabela, M. L. 2015. Integration of agroindustrial co-products as functional food ingredients: Cactus pear (*Opuntia ficus indica*) flour and pineapple (*Ananas comosus*) peel flour as fiber source in cooked sausages inoculated with lactic acid bacteria. *Journal of Food Processing and Preservation* 39: 2630-2638.

Dibanda, R. F., Rani, P. A., and Manohar, R. S. 2016. Chemical composition of some selected fruit peels. *European Journal of Food Science and Technology* 4: 12-21.

Emaga, T. H., Herinavalona, A. R., Wathelet, B., Tchango, T. J., and Paquot, M. 2007. Effects of the stage of maturation and varieties on the chemical composition of banana and plantain peels. *Food Chemistry* 103: 590-600.

Ferreira, E. A., Siqueira, H. E., Vilas Boas, E. V., Hermes, V. S., and Oliveira, R. A. 2016. Bioactive compounds and antioxidant activity of pineapple fruit of different cultivars. *Revista Brasileira de Fruticultura* 38: 1-7.

Fisher, J. 2012. Fruit Fibers. In: *Dietary Fibers and Health*. Cho, S. S., and Almeida, N. (Eds.) CRC Press, Boca Raton, Florida, pp. 409-420.

Gibson, G. R., and Roberfroid, M. B. 1995. Dietary modulation of the colonic microbiota: Introducing the concept of prebiotics, *Journal of Nutrition*, 125: 1401-1412.

Hemalatha, R., and Anbuselvi, S. 2013. Physicochemical constituents of pineapple pulp and waste. *Journal of Chemical and Pharmaceutical Research* 5: 240-242.

Howlett, J. 2008. Functional Foods. From Science to health and claims. *ILSI Europe Concise Monograph Series*. Aggett, P. (Ed.) Belgic. ISBN 9789078637110.

Howlett, J. F., Betteridge, V. A., Champs, M., Craig, S. A. S., Meheust, A., and Miller, J. J. 2012. Discussions relating to the definition of dietary fiber at the ninth Vahouny fiber Symposium. In: *Dietary Fibers and Health.* Cho, S. S., and Almeida, N. (Eds.) CRC Press, Boca Raton, Florida, pp. 11-18.

Huang, Y-L., Chow, Ch-J., and Fang, Y-J. 2011. Preparation and physicochemical properties of fiber rich fraction from pineapple peels as a potential ingredient. *Journal of Food and Drug analysis* 19: 318-323.

Huang, Y-L., Tsai, Y-H., and Chow, C-J. 2014. Water-insoluble-fiber-rich fraction from pineapple improves intestinal function in hamsters: evidence from cecal and fecal indicators. *Nutrition Research 34*: 346-354.

Kubomara, K. 1998. Japan redefines functional foods. *Prepared Foods 167*: 129-132.

Larrauri, J. A., Rúperez, P., and Saura-Calixto, F. 1997. Pineapple Shell as a source of dietary fiber with associated polyphenols, *Journal of Agriculture and Food Chemistry* 45: 4028-4031.

Li, T., Shen, P., Liu, W., Liu, Ch., Liang, R., and Chen, N. Y. 2014. Major polyphenolics in pineapple peels and their antioxidant interactions. *International Journal of Food Properties* 17: 1805-1817.

Martínez, R., Torres, P., Meneses, M. A., Figueroa, J. G., Pérez-Álvarez, J. A., and Viuda-Martos, M. 2012. Chemical, technological and *in vitro* antioxidant properties of mango, guava, pineapple and passion fruit dietary fiber concentrate. *Food Chemistry* 135: 1520-1526.

Mehta, N., Ahlawat, S. S., Sharma, D. P., and Dabur, R. S. 2015. Novel trends in development of dietary fiber rich meat products-a critical review. *Journal of Food Science and Technology* 52: 633-647.

Morais, D. R., Rotta, E. M., Sargi, S. C., Bonafe, E. G., Suzuki, R. M., Souza, N. E., Matsushita, M., and Visentainer, J. R. 2017. Proximate composition, mineral contents and fatty acid composition of the different parts and dry peels of tropical fruits cultivated in Brazil. *Journal of the Brazilian Chemical Society*, 28: 308-318.

Novaes, L. C., Jozala, A. F., Lopes, A. M., Santos-Ebinuma, V. C., Mazzola, P. G., and Pessoa, J. A. 2016. Stability, purification and applications of bromelain: A Review. *Biotechnology Progress* 32: 5-13.

Olmedilla-Alonso, B., Jiménez-Colmenero, F., and Sánchez-Muniz, F. J. 2013. Development and assessment of healthy properties of meat and meat products designed as functional foods. *Meat Science* 95: 919-930.

Parra-Matadamas, A., Mayorga-Reyes, L., and Pérez-Chabela, M. L. 2015. *In vitro* fermentation of agroindustrial by-products: grapefruit albedo and peel, cactus pear peel and pineapple peel by lactic acid bacteria. *International Food Research Journal* 22: 859-865.

Pogorzelska-Nowicka, E., Atanasov, A. G., Horbanczuk, J., and Wierzbicka, A. 2018. Bioactive compounds in functional meat products. *Molecules* 23: 307-326.

Pyar, H., Liong, M-T., and Peh, K. K. 2014. Potentials of pineapple waste as growth medium for Lactobacillus species. *International Journal of Pharmacy and Pharmaceutical Sciences* 6: 142-145.

Ramírez, A., and Pacheco de Delahaye, E. 2011. Chemical composition and bioactive compounds in pineapple, guava and soursop pulp. *Interciencia* 36: 71-75.

Rashad, M. M., Mahmoud, A. E., Ali, M. M., Nooman, M. V., and Al-Kashed, A. S. 2015. Antioxidant and anticancer agents produced from pineapple waste by solid state fermentation. *International Journal of Toxicological and Pharmacological Research* 7: 287-296.

Ribeiro da Silva, L. M., Teixeira de Figueiredo, E. A., Pontes Silva, R. N. M., Pinto Vieira, I. G., Wilane de Figueiredo, R., Montenegro, B. I., and Gomes, C. L. 2014. Quantification of bioactive compounds in pulps and by-products of tropical fruits from Brazil. *Food Chemistry* 143: 398-404.

Roberfroid, M. 2007. Prebiotics: the concept revisited. *Journal of Nutrition* 137: 830S-837S.

Sah, B. N. P., Vasilejic, T., McKechnie, and Donko, O. N. 2016. Physicochemical, textural and rheological properties of probiotic yoghurt fortified with fibre-rich pineapple peel powder during refrigerated storage. *LWT-Food Science and Technology* 65: 978-986.

Sánchez Pardo, M. E., Ramos Cassellis, M. E., Mora Escobedo, R., and Jiménez García, E. 2014. Chemical characterisation of the industrial residues of the pineapple (*Ananas comosus*). *Journal of Agricultura Chemistry and Environment* 3: 53-56.

Saraswaty. V., Risdian, C., Primadona, I., Andriyani, R., Andayani, D. G. S., and Mozef, T. 2017. Pineapple peel wastes as a potential source of antioxidant compounds. *IOP Conf. Series: Earth and Environmental Science* 60: 1-6.

Selani, M. M., Shirado, G. A. N, Margiotta, G. B., Saldaña, E., Spada, F. P., Piedade, S. M. S., Contreras-Castillo, C. J., and Canniatti-Brazaca, S.G. 2016b. Effects of pineapple byproduct and canola oil as fat replacers on physicochemical and sensory qualities of low-fat beef burger. *Meat Science* 112: 69-76.

Selani, M. M., Shirado, G. A. N., Margiotta, G. B., Rasera, M. L., Marabesi, A. C., Piedade, S. M. S., Contreras-Castillo, C. J, and Canniatti-Brazaca, S. G. 2016a. Pineapple by-product and canola oil as partial fat replacers in low-fat beef burger: Effects on oxidative stability, cholesterol content and fatty acid profile. *Meat Science* 115: 9-15.

St. Clair Henning S, P Tshalibe, LC. Hoffman Physico-chemical properties of reduced-fat beef species sausage with pork back fat replaced by pineapple dietary fibres and water. *LWT - Food Science and Technology* 74 (2016) 92e98.

Steingass, C. B., Glock, M.P., Schwiggert, R. M., and Carle, R. 2015. Studies into the phenolic patters of different tissues of pineapple (*Ananas comosus (L.) Merr.*) infructescence by HPLC-DAD-ESI-MS and GC-MS analysis. *Analytical and Bioanalytical Chemistry* 407: 6463-6479.

Teixeira-Souza, R. A., Branco da Fonseca, T. R., de Souza Kirsch, L., Cavalcante, S. L. S., Alecrim, M. M., da Cruz, F. R. F., and Simas, T. M. F. 2016. Nutritional composition of bio products generated from semi-solid fermentation of pineapple peel by edible mushrooms. *African Journal of Biotechnology* 15: 451-457.

Trowell, H. 1977. Food and Dietary fiber. *Nutrition Reviews*: 6-11.

BIOGRAPHICAL SKETCHES

Alfonso Totosaus

Affiliation: Food Science Lab & Pilot Plant, Tecnologico Estudios Superiores Ecatepec.

Education: PhD in Biological Sciences.

Research and Professional Experience: Research in functional properties of food components and food texture.

Professional Appointments: National Researchers System level 2.

Honors: Nothing to mention yet.

Publications from the Last 3 Years:

Pintor-Jardines A., Arjona-Román, J. L., Totosaus-Sánchez, A., Severiano-Pérez, P., González-González, L. R., Escalona-Buendia, H. B. 2018. The influence of agave fructans on thermal properties of low-fat, and low-fat and sugar ice cream. *LWT-Food Science and Technology* 93: 679-685. DOI: 10.1016/j.lwt.2018.03.060.

Serrano-Casas V, Pérez-Chabela, M. L., Cortés-Barberena, E., Totosaus, A. 2017. Improvement of lactic acid bacteria viability in acid conditions employing agroindustrial co-products as prebiotic on alginate ionotropic gel matrix co-encapsulation. *Journal of Functional Foods* 38: 293-297. DOI: 10.1016/j.jff.2017.09.048.

Díaz-Vela, J., Totosaus, A., Escalona-Buendía, H. B., Pérez-Chabela, M. L., 2017. Influence of the fiber from agro-industrial co-products as functional food ingredient on the acceptance, neophobia and sensory characteristics of cooked sausages. *Journal of Food Science and Technology* 54(2): 379-385. DOI: 10.1007/s13197-016-2473-8.

Téllez-Rangel, E. C., García-Martínez, I., Rodríguez-Huezo, E., Totosaus, A., 2017. Películas comestibles de proteína con aceite de maíz emulsionado como acarreador de paprika para mejorar color y oxidación de lípidos en Boloña de res. *Nacameh* 11 (1): 18-33. [Edible protein films with corn oil emulsified as a paprika carrier to improve color and oxidation of lipids in beef Bolognese.]

Hernández-Alcántara, A. M., Totosaus, A., Pérez-Chabela, M. L., 2016. Evaluation of agro-industrial co-products as source of bioactive compounds: fiber, antioxidants and prebiotic. *Acta Universitatis Cibiniensis. Series E Food Technology* 20 (2): 3-16, DOI: 10.1515/aucft-2016-0011.

Fragoso, M. M., Pérez-Chabela, L., Hernández-Alcántara, A. M., Escalona-Buendía, H. B., Pintor, A., Totosaus, A., 2106. Sensory, melting and textural properties of fat-reduced ice cream inoculated with thermotolerant lactic acid bacteria. *Carpathian Journal of Food Science and Technology* 8(2): 11-21.

Perez-Rocha, K. A., Guemes-Vera, N., Bernardino-Nicanor, A., Gonzalez-Cruz, L., Hernandez-Uribe, J. P., Totosaus Sanchez, A., 2015. Fortification of white bread with guava seed protein isolate. *Pakistan Journal of Nutrition*, 14 (11): 828-833. DOI: 10.3923/pjn.2015-828-833.

Díaz-Vela, J., Totosaus, A., Pérez-Chabela, M. L., 2015. Integration of agroindustrial by-products as functional food ingredients: cactus pear (*Opuntia ficus indica*) flour and pineapple (*Ananas comosus*) peel flour as fiber source in cooked sausages inoculated with lactic acid bacteria. *Journal of Food Processing and Preservation*, 39(6): 2630-26358. DOI: 10.1111/jfpp.12513.

Rojas-Nery, E., Güemes-Vera, N., Meza-Márquez, G. O., Totosaus, A., 2015. Carrageenan type effect on soybean oil/soy protein isolate emulsion employed as fat replacer in panela-type cheese. *Grasas Y Aceites*, 66 (4): e097. DOI: 10.3989/gya.0240151.

Castillejos-Gómez, B. I., Totosaus, A., Pérez-Chabela, M. L., 2015. Use of maguey (Agave spp.) leaves as a source of functional ingredients in cooked sausages elaboration. In: *Advances in Science, Biotechnology and Safety of Foods*. Santos García, Hugo Sergio García Galindo & Guadalupe Virginia Nevárez-Moorillón (Editors). Asociación Mexicana de Ciencias de los Alimentos, A. C., Durango, ISBN: 978-607-95455-4-3, pp. 163-170.

Rojas-Nery, E., Garcia-Martinez, I., Totosaus, A., 2015. Effect of emulsified soy oil with different carrageenans in rennet-coagulated milk gels. *International Food Research Journal* 22(2): 606-612.

Güemes Vera, N., Totosaus Sánchez, A., 2015. Mechanical properties of White Bread made of barley flour concentrate and *Jatropha curcas*. In: *Jatropha Curcas Biology, Cultivation and Potential Uses*. G. Medina (Editor). ISBN 978-1-63483-112-3, Nova Science Publishers, Inc., Hauppauge, pp. 203-213.

Pérez-Chabela, M. L., Chaparro-Hernández, J., Totosaus, A., 2015. Dietary fiber from agroindustrial by-products: orange peel flour as functional ingredient in meat products. In: *Dietary Fiber: Production Challenges, Food Sources and Health Benefits,* Marvin E. Clemens (editor). ISBN 978-1-63463-676-6, Nova Science Publishers, Inc., Hauppauge, pp. 145-157.

Totosaus, A, González-González, R., Fragoso, M. 2016. Influence of the type of cellulosic derivatives on texture, oxidative and thermal stability of soybean oil oleogel. *Grasas Y Aceites*, 67 (3). DOI: 10.3989/gya.0440161.

Hernández-García, S., Salazar-Montoya, J. A., Totosaus, A., 2016. Emulsifying properties of food proteins conjugated with glucose or lactose by two methods (spray-drying or freeze-drying). *International Journal of Food Properties*, 19(3): 526-536. DOI: 10.1080/10942912.2015.1033551.

Totosaus A., Ariza-Ortega, T. J., 2016. Carne y productos cárnicos como fuente de péptidos bio-activos. *Nacameh* 10 (2): 49-58. [Meat and meat products as a source of bio-active peptides.]

Pintor, A., Escalona-Buendia, H. B., Totosaus, A. 2017. Effect of inulin on melting and textural properties of low-fat and sugar-reduced ice cream: optimization via a response surface methodology. *International Food Research Journal* 24(4): 1728-1734.

Totosaus, A., Rojas-Nery, E., Franco-Fernández, M. J., 2017. Soya bean oil/soy protein isolate and carrageenan emulsions as fat replacer in fat-reduced Oaxaca-type cheese. *International Journal of Dairy Technology* 70(4): 499-505. DOI: 10.1111/1471-0307.12397.

Bazán Lugo, E., Álvarez Cárdenas, C. G., Totosaus A., S. 2017. Efecto del pretratamiento de la paja de trigo sobre el rendimiento de biomasa y la producción de un extracto con actividad celulolítica empleando *Pseudomonas aeruginosa*. *Acta Universitaria* 27(5): 26-33.DOI: 10.15174/au.2017.1276. [Effect of pretreatment of wheat straw on biomass yield and production of an extract with cellulolytic activity using *Pseudomonas aeruginosa*.]

Téllez-Rangel, E. C., Rodríguez-Huezo, E., Totosaus, A. 2018. Effect of gellan, xanthan or locust bean gum and/or emulsified maize oil on proteins edible films properties. *Emirates Journal of Food and Agriculture* 30(5): 404-412. DOI: 10.9755/ejfa.2018.v30.i5.1644.

M. Lourdes Pérez-Chabela

Affiliation: Biotechnology Department, Universidad Autonoma Metropolitana

Education: PhD in Biological Sciences

Professional Appointments: National Researchers System level 2

Publications from the Last 3 Years:

Díaz-Vela, J., Totosaus, A., Escalona, H., Pérez-Chabela, M. L. 2017. Influence of the fiber from agro-industrial co-products as functional food ingredient on the acceptance, neophobia and sensory characteristics of cooked sausages. *Journal of Food Science and Technology*, 54:379-385.

Fragoso, M., Pérez-Chabela, M. L., Hernández-Alcántara, A. M., Escalona-Buendía, H. B., Pintor, A., Totosaus, A. 2016. Sensory melting and textural properties of fat-reduce ice cream inoculated with thermotolerant lactic acid bacteria, *Carpathian Journal of Food Science and Technology*, 8(2): 11-21.

Hernández-Alcántara, A. M., Totosaus, A., Pérez-Chabela, M. L. 2016. Evaluation of agro-industrial co-products as source of bioactive compounds: fiber, antioxidants and prebiotic. *Acta Universitatis Cibinensis Series E: Food Tecnology*, 20(2): 3-16.

Hernández-Alcántara, A. M., Wacher, C., Goretti Llamas, M., López, P., Pérez-Chabela, M. L. 2018. Probiotic properties and stress response of thermotolerant lactic acid bacteria isolated from cooked meat products, *LWT_ Food Science and technology*, 91:249-257.

Kaktcham, P. M., Bocamdé Temgoua, J., Ngoufack Zambou, F., Díaz-Ruiz, G., Wacher, C., Pérez-Chabela, M. L. 2018. *In vitro* evaluation of the probiotic and safety properties of bacteriocinogenic and non-bacteriocinogenic lactic acid bacteria from the intestines of Nile tilapia and common carp for their use as probiotics in aquaculture. 2017. *Probiotic and Antimicrobial proteins*, 10(11): 98-109.

Kaktcham, P. M., Bocamdé Temgoua, J., Ngoufack, Zambou, F., Díaz-Ruíz, G., Wacher, C., Pérez-Chabela, M. L. 2016. Quantitative analysis of the bacterial microbiota of rearing environment, tilapia and common carp cultures in earthen ponds and inhibitory activity of its lactic acid bacteria on fish spoilage and pathogenic bacteria. *World Journal of Micorbiology Biotechnology*, 33: 32.

Serrano-Casas, V., Pérez-Chabela, M. L., Cortes-Barberena, E., Totosaus, A. 2017. Improvement of lactic acid bacteria viability in acid conditions employing agroindustrial co-products as prebiotic on alginate ionotropic gel matrix co-encapsulation. *Journal of Functional Foods,* 38:293-297.

INDEX

D

E

F

G

H

I

K

L

M

N

O

P

W

β